T0340014

# Substance Abuse
## *in the*
# Workplace

## Reginald L. Campbell
## R. Everett Langford

**CRC Press**
Taylor & Francis Group
Boca Raton  London  New York

CRC Press is an imprint of the
Taylor & Francis Group, an **informa** business

CRC Press
Taylor & Francis Group
6000 Broken Sound Parkway NW, Suite 300
Boca Raton, FL 33487-2742

First issued in paperback 2020

ISBN-13: 978-0-367-44909-4 (pbk)
ISBN-13: 978-0-87371-131-9 (hbk)

This book contains information obtained from authentic and highly regarded sources. Reasonable efforts have been made to publish reliable data and information, but the author and publisher cannot assume responsibility for the validity of all materials or the consequences of their use. The authors and publishers have attempted to trace the copyright holders of all material reproduced in this publication and apologize to copyright holders if permission to publish in this form has not been obtained. If any copyright material has not been acknowledged please write and let us know so we may rectify in any future reprint.

For permission to photocopy or use material electronically from this work, please access www. copyright.com (http://www.copyright.com/) or contact the Copyright Clearance Center, Inc. (CCC), 222 Rosewood Drive, Danvers, MA 01923, 978-750-8400. CCC is a not-for-profit organiza- tion that provides licenses and registration for a variety of users. For organizations that have been granted a photocopy license by the CCC, a separate system of payment has been arranged.

**Trademark Notice:** Product or corporate names may be trademarks or registered trademarks, and are used only for identification and explanation without intent to infringe.

**Visit the Taylor & Francis Web site at
http://www.taylorandfrancis.com**

**and the CRC Press Web site at
http://www.crcpress.com**

## Library of Congress Cataloging-in-Publication Data

Catalog record is available from the Library of Congress

# ■ PREFACE

It is hoped that this book will help employers and employees comply with the Drug-Free Workplace Act as well as to help in stemming the tide of drug abuse sweeping the nation at this time.

Our appreciation is expressed to many people — Jon Lewis of Lewis Publishers, the editors at CRC Press, especially Helen Linna, as well as the students we tested this material on.

Both of us would like to express deep appreciation for the understanding of the ladies in our lives, Reneé Scudder and Cecilia Son-Hee Langford, during the many hours in front of the word processor away from them. This book is dedicated to them.

To all others who have helped us in this endeavor, thank you.

Any errors, omissions, and other mistakes belong to us.

**Reginald L. Campbell**
Sierra Vista, Arizona
**R. Everett Langford**
Dayton, Ohio

# THE AUTHORS

**Mr. Campbell** was born in Hartford, Connecticut. His undergraduate degree was from Fairmont State College, Fairmont, West Virginia, and Yale-New Haven Hospital, New Haven, Connecticut. He received his M.S. from Marshall University, Huntington, West Virginia. He is currently completing requirements for a Ph.D. in ergonomics and safety. Mr. Campbell has attended a number of educational and training courses in drug abuse, chemical substances, industrial hygiene, and hazardous materials management.

He was an Assistant Professor of Medicine, Hahnemann Medical College of Allied Health Sciences, Philadelphia, and Instructor of Occupational Health at the U.S. Department of Labor's National Mine Health and Safety Academy, Beckley, West Virginia. He has served as Industrial Hygienist for the U.S. Army Medical Department Activity, Fort Huachuca, Arizona.

He is presently Chief Executive Officer of Campbell Associates, Inc., Sierra Vista, Arizona, a consulting firm specializing in safety and health issues. Campbell Associates, Inc. has assisted a number of companies, governmental agencies, and labor unions in matters relating to hazard communication, hazardous materials management, and drug abuse education.

Mr. Campbell is co-author of *Fundamentals of Hazardous Materials Incidents* (Lewis Publishers, 1991) and has published numerous articles on substance abuse in the workplace. He is noted in a number of biographical references, including *Who's Who in the Southwest, Who's Who in the World,* and *Personalities of America.*

**Dr. Langford** was born in Owensboro, Kentucky and grew up in Savannah, Georgia. He received an A.A. from Armstrong State College in 1965, a B.S. in chemistry and physics from Georgia Southern College in 1967 and an M.S. in solution thermodynamics in 1971 and a Ph.D. in physical geochemistry in 1974, both from the University of Georgia under a National Defense Education Act Fellowship.

He taught chemistry, geology, environmental chemistry, and physical science at the Savannah Science Museum, Bainbridge College, Georgia Military College, and Georgia Southern College before becoming the Chief of Clinical Chemistry at the U.S. Army Academy of Health Sciences. Following that assignment, Dr. Langford served as a consultant in hazardous waste management at the U.S. Army Environmental Hygiene Agency and then as Commander of the Environmental Sanitation Detachment in Taegu, Korea for the U.S. Army's 5th Preventive Medicine Unit. In Korea, he served as an adjunct faculty member to the

University of Maryland. Upon his return to the U.S. he became Chief of Environmental Health at Fort Huachuca, Arizona and Preventive Medicine Officer for the U.S. Army Information Systems Command, followed by an assignment as Commander, U.S. Army Medical Research Detachment at Wright-Patterson Air Force Base, Dayton, Ohio.

Dr. Langford has served as a Judge at the Westinghouse International Science Fair, a member of the review panel for the National Science Foundation's Comprehensive Assistance to Undergraduate Science Education, a member of the scientific review panel for the Hazardous Substances Data Bank of the National Library of Medicine, and a consultant to the International Union of Operating Engineers. He is a Certified Industrial Hygienist (CIH), Registered Hazardous Substances Professional (RHSP), Registered Sanitarian (RS), Fellow of the American Institute of Chemists (FAIC), Diplomate of the American Academy of Sanitarians, and Engineer-in-Training (EIT). He is author or co-author of some twenty papers in thermodynamics, geochemistry, and hazardous materials, and is co-author of *Fundamentals of Hazardous Materials Incidents*. He is listed in some fifteen biographical references, including *Who's Who in the World, Who's Who in Science and Engineering,* and *Who's Who of Emerging Leaders.*

Dr. Langford also completed three years of postgraduate study at the University of North Carolina - Chapel Hill in radiological hygiene.

# CONTENTS

# ■ INTRODUCTION

# 1

## ■ 1.1. WHO IS THIS BOOK FOR?

There are several audiences for which this book is intended: middle and upper management, labor leaders, industrial hygiene and safety personnel, and workers. The management team is the primary target audience, since they can benefit from the information about setting up Employee Assistance Programs, the legal requirements of employers and employees, the effects on their business, and the documentation available in the appendices. Labor leaders can use the information on sources of help available and what the effects of substance misuse can be to work with management in solving the problem. Industrial hygiene and safety officers often have the closest daily contact with the largest number of workers and can see the effects of substance abuse first; in addition, it is important for these people to know the effects, legal requirements, and sources of assistance available to help the workers they are sworn to protect. Finally, all workers need to understand the effects of drugs on the functions of the body as well as sources of help when substance abuse occurs. Each of these groups, and others, may select the chapters of most interest to them and omit others as time allows. In addition, it is hoped that this book can be a valuable asset to the company training department by putting a lot of information into one place. In fact, the book, used as a reference, could become the basis of a training and awareness program within the company.

## ■ 1.2. IMPORTANCE OF THE PROBLEM

Most people will agree that substance abuse, the nonmedical use of drugs (especially alcohol), is today a worldwide problem. We have only to read a newspaper or news magazine; listen to the radio or television; attend a school conference or PTA meeting; listen to law enforcement, health, and school officials; or just talk to our children to grasp

1

the magnitude of the problem. Yet, with millions of dollars already spent and currently being expended, countless hours of education and training, and warehouses of literature produced, the substance abuse problem continues to grow — reaching into almost every area of daily life and leaving virtually no one untouched, regardless of age, socioeconomic condition, or occupation. This is particularly true in the workplace.

Substance abuse can adversely affect employees' physical and emotional health as well as social life and job performance. The U.S. Department of Health and Human Services (HHS) estimates that one worker in five throughout the nation's workplaces suffers from personal problems stemming from drug or alcohol abuse, or emotional problems. With so many drug abusers in the country today, you can bet you have some on the plant payroll and, at least some of the time, they're in no shape to work. This translates into excessive absenteeism, vulnerability to accidents, and shoddy job performance. The U.S. Department of Justice's Drug Enforcement Administration (DEA) advises that the epidemic of drug abuse that swept the nation in the late 1960s and 1970s is being felt in the workplace, with managers beginning to realize the drug abuse problem can result in absenteeism; tardiness; loss of productivity and efficiency; increased accidents, liability and health costs; and theft and security problems. Definitively, the DEA reports: "Although alcohol is still the nation's most serious drug problem, use of other drugs is prevalent throughout the industrial sector and affects both large and small companies."

The substance abuser is not only the heroin addict or the habitual drunkard; in fact, these are relatively small numbers of the problem. The real problem is the millions who abuse alcohol, marijuana, cocaine, etc., with the result that they are not quite "up to speed". Workers who are "mellowed" or "keyed up" on drugs or hung over with alcohol or suffering from "flashbacks" are not able to perform their jobs to the best of their ability. They may not even consider themselves addicts or abusers, yet they daily cause reduced efficiency and productivity in our factories and shops and offices; they cause serious accidents or even deaths to happen to themselves or other workers. These workers simply aren't at their best when affected by chemical substances. In addition, substance misuse also affects employees' families and reduces the employee's self respect, with the result that some of your workers have created additional problems for themselves — and for you.

If we can agree on the magnitude of the substance abuse problem, we can also admit it is pervasive enough to warrant our direct attention as a serious issue in the workplace — especially your workplace — and take steps to generate preventive and corrective measures to deal not only with the health and safety of the substance abuser, but also with the health and safety of those who work in the substance abuser's environment!

## ■ 1.3. PARADIGMS AND POLITICS

There are several problems inherent in discussing substance abuse and substance abusers: emotional difficulties (including discussions of morality) and legal difficulties.

The current paradigm or philosophical concept requires that one accept the notions that (1) drug abuse is wrong and (2) some people abuse drugs; therefore, (3) drug abusers are wrong and should be punished. Unfortunately, this idea has not reduced the prevalence of drug misuse in this nation and may have even contributed to it. Of course, chemical misuse is wrong, and there are people who abuse drugs, but it does not necessarily follow that such people need to be punished. Most abusers, when not under a craving for their chemical of choice, want out of their cycle of ups and downs and dependency on chemicals; the problem lies in how to help them do just that.

Law enforcement activities have largely centered on the street user of drugs because the user is often forced to steal or turn to prostitution in order to pay for the habit and are thus easy targets for the police. Catching and convicting major drug distributors is extremely difficult since they are usually clever and wealthy enough to avoid being caught. Interestingly, most major drug dealers never use drugs and forbid their children to do so. A large number of arrests may suggest that the police are "doing something" about the problem, but such "busts" rarely catch anyone higher than a street dealer. The police are in a no-win situation: the public demands arrests while the real kingpins continue to evade arrest. Many police are frustrated by the lack of progress in removing drugs from the streets and homes of America. Arrests simply do not work! The only way that drug misuse will decrease is for the users and potential users to realize the dangers and voluntarily not get involved in the first place or seek treatment if they have started. *Thus, education is the only answer.*

The casual user of illicit substances may feel that he or she isn't a part of the criminal underground, but they are the source of money to the drug lords. Even the most casual user is channeling money into a variety of criminal and even terrorist activities without thinking about it.

The question of morality is valid, although most people don't like being preached at or lectured to. Simply letting the users know that they are buying more than just the drug is important but difficult to do. The criminal underground thrives on the drug trade much as the gangsters of the 1920s thrived on illegal alcohol; the profits are used to enter into ever-widening areas of criminal activity. Several noted terrorist organizations have ties to the drug trade; the drug user is supporting anti-American terrorism in making his or her purchase of drugs here.

We have to create a new paradigm, one in which people themselves don't want and have no need for nor desire to abuse drugs; therefore, education is the answer. If people understand what chemicals can do to their bodies and the terrible results of chemical substance abuse, perhaps they will voluntarily decide that the price is too high for the brief feelings of pleasure which result from the chemicals coursing through their blood. Elimination of hopelessness and despair together with increased awareness of the harmful effects of drug misuse would do more for the ending of chemical abuse than all the arrests and after-the-fact treatments we see today. Hopefully, politics will allow this to happen.

## ■ 1.4. PURPOSE OF THIS BOOK

This book can be a valuable part of a program to comply with the Drug-Free Workplace Act of 1988, which requires that any company receiving a grant or contract from the Federal government in the amount of $25,000 or more establish a program to educate all employees as to the dangers of drug abuse in the workplace. It is the intention of this book to inform the reader, who may be a supervisor, manager, employee, or co-worker of a person who abuses drugs, to better understand the reasons, physiological as well as emotional, for drug abuse and to help the person suffering from this progressive abuse. To this end, the book begins the history of substance abuse, then discusses how the human body functions normally or under the influence of chemicals, followed by a toxicological description of the more common chemicals abused today in America. The book then discusses ways of helping the abuser through identification and assistance programs available. Since many of the drugs are regulated or controlled, a chapter will cover the laws involved. Specific work places will be discussed with some case histories as well as typical drugs used and concerns specific to that profession or trade.

It is not the purpose of this book to make anyone a trained toxicologist, biologist, chemist, law enforcement official, or even drug assistance counselor; these fields require years of extensive training and experience. It is, however, the purpose of this book to make people involved with drug abuse in the workplace aware of the problem, symptoms of the abuser, and ways to help at their level. The authors are sincerely concerned with the epidemic of drug abuse and the crime it spawns currently sweeping this country; however, they are reminded that, in the 1920s, gangsters rode rampant over the streets and byways of America while it appeared that nothing could be done; yet, this era came to an end when dedicated people from many walks of life decided that they had enough. The same can be true for drug abuse and chemical dependency today!

# 2 BRIEF HISTORY OF SUBSTANCE ABUSE

## 2.1. INTRODUCTION

Many people think that the beginning of substance abuse and chemical dependency in this country was the "hippie" movement of the 1960s or perhaps the cocaine craze of the 1920s. These assumptions are incorrect; substance abuse began as soon as humans realized that certain leaves, twigs, cacti, barks, etc. could be used to produce an altered state of consciousness or relief from pain. History books, historical newspapers, and even the Bible contain references to chemical substances being used to alter perception or reality. Early people used plants and herbs to cure or relieve the maladies that they experienced. Herbs and plants were the basis for all the early "medicines". Many of these materials did, indeed, have medicinal properties. For example, early Native Americans used the bark of the willow tree to make a tea to give relief from body aches and pains as well as headaches; we now know that willow bark contains a form of aspirin. Some substances were addictive or capable of producing hallucinations. These hallucinations were thought by some early peoples to be communication with the spirit world. Thus, not all the "medicines" were used to seek relief from pain; some were used to seek escape from reality. Today is no different.

People have been using "natural" drugs for centuries. Almost all prescriptions or home remedies were concocted from materials present in nature, not produced on the chemist's bench. This was true until the introduction of synthetic chemicals at the beginning of the 1900s.

## 2.2. CHEMICAL DEPENDENCY AND SUBSTANCE ABUSE

What are substance abuse and chemical dependency? These must be defined in order to understand the scope of the problem, how the problem

5

originated, and what the impact is on our society, in our homes, and in our workplaces. A panel of the United Nations defined **substance abuse** as, "A state of periodic or chronic intoxication detrimental to the individual and society, produced by the repeated consumption of natural or synthetic drugs."

Substance abuse and chemical dependency could probably be best described as addiction to any chemical. This could be nicotine in cigarettes, caffeine in coffee or cocoa products, or even tomatine in tomatoes, or to prescription medications ordered by a physician. All of these have addictive properties; they are capable of "periodic or chronic intoxication detrimental to the individual and society...." Later chapters will go into greater detail about chemicals of abuse, chemical dependency, and the ways in which employers and fellow workers can be of help to the abuser.

## ▩ 2.3 EARLY HISTORY

Early people used the plants and herbs that were found to prepare salves, ointments, teas, and potions. Some worked; many did not; some even killed. By trial and error, early people experimented with natural substances to find which ones helped cure pain and illness. These early "chemists" noted that by using marijuana, coffee, tea, morning-glory seeds, peyote, certain mushrooms, etc., they could have a feeling of stimulation, of being detached from the body, seemingly able to communicate with their gods, or able to see strange images.

Chinese writings from 2700 B.C. describe a plant, *ma huang*, as being able to revive and stimulate; today, we know that this plant contains ephedrine, a stimulant. Socrates died from drinking oil of the hemlock plant as a sentence for his teachings suspected of corrupting the young; but, in smaller doses, it could be used to relieve pain. The Chinese and Koreans have eaten the root of the ginseng plant as a tonic and supposed aphrodisiac for thousands of years. Native Americans have used peyote and mescal in religious ceremonies for generations. Coca leaves have been chewed for their sedative effects, and poppy extracts have been used to relieve pain. South American tribes have used natural drugs to stun and kill prey for hundreds of years.

Ethyl alcohol has been known for thousands of years; its name comes from the Arabic *al-koh'l* meaning "subtle". Most peoples of the world have found a way to ferment grains, roots, or even milk into alcohol-containing drinks.

Early peoples found many plant and animal substances which could produce a medical effect: **quinine** from the bark of the cinchona tree of South America; **cocaine** from the coca plant; **strychnine** and **curare** from strychnos plants of the West Indies and the Philippines; **nicotine** from the tobacco plant; **codeine** and **morphine** from the opium poppy; **ricinine** from the castor bean; **atropine** from the deadly nightshade; **digitalis** from the common foxglove plant; **reserpine** from the snake root plant; **physostigmine** from the calabar plant; and **caffeine** from the coffee plant. There are hundreds of others, most of which possess some physiological effects.

There were also hallucinogenic substances such as ibogaine, peyote, and mescaline discovered. Certain mushrooms contain hallucinogenic substances, the most common being the sacred mushroom of Mexico which contains chemicals capable of causing sensory alterations. Some cacti contain natural hallucinogens, the most common being the peyote cactus; its mescal buttons contain mescaline, a powerful hallucinogen.

Marijuana was discovered by early peoples of Mexico and the desert southwest and elsewhere in the world. It consists of leaves, stems, and flowering tops of the hemp plant.

## ▬ 2.4 OPIUM

The earliest recorded use of opium was by the Sumerians, who, about the year 4000 B.C., extracted it from poppy plants. The Ebers papyrus, circa 1500 B.C., makes reference to a remedy "to prevent the excessive crying of children." This property of opium was later used in the patent medicine paregoric.

Around the year 700 B.C., the Egyptians adopted in their hieroglyphics the same symbol that had appeared on Sumerian tablets along with a description on how to harvest the opium plant and collect the sap.

Opium was one of the most frequently traded items in the Middle Ages, being obtained by Arab merchants in regions of the Middle and Far East and transported to Europe.

The Greek physician Erasistratrus warned of the addictive power of opium. Galen, the last of the great Greek physicians, felt that opium could be used as a cure for most diseases and mentioned how opium cakes and candies were being sold in the market place. Homer's *Odyssey*, circa 1000 B.C., notes that a drug called *nepenthes* gave forgetfulness of sorrow. Some authors are sure that this refers to opium. Greek and Roman mythology refer to the gods of sleep as Hypnos and Somnus. Both were usually depicted adorned with or carrying poppies or an opium container. Somnus was frequently depicted as pouring juice from an opium container into the eyes of a sleeping person. J. M. Scott in his book, *The White Poppy: A History Of Opium*, mentions that the Roman Emperor Marcus Aurelius clearly indicated the signs and symptoms of being addicted to opium and demonstrated withdrawal symptoms. Hippocrates, the "father of medicine", was known to prescribe medicines made from local herbs, including opium. Greek knowledge of the use of medicines using opium declined with the fall of the Roman Empire. Following this decline there was not much use of opium in the Western world for the next 1000 years.

Chinese literature first refers to the opium poppy around the years 600–700 A.D. According to Chinese legend, Buddha cut off his eyelids to keep from falling asleep and as they fell to the ground, poppy plants sprang from the earth.

The Arab world of northern Africa used opium and hashish because they could not use alcohol as prohibited by the Koran. Opium and hashish became the major social drugs of the Islamic faith as the religion spread across the Middle East. During the Dark Ages in Europe, the Arab world was expanding greatly, occupying much of Spain and all of northern Africa. With this expansion, opium use also expanded.

## ■ 2.5. COCA AND COCAINE

Many early people looked for substances that would give a sense of well being and exhilaration. The Incas of South America discovered such a material — the leaves of the coca plant. The Incas lived in the high Andes Mountains and developed a very sophisticated communications system throughout their empire which was dependent upon couriers. In order for the couriers to run the long distances between cities, endure the very thin atmosphere of the high altitudes (10,000 to 18,000 feet above sea level), and be able to run on limited food intake, they chewed coca leaves to sustain the energy loss and pain necessary to run these distances. The Incas and their descendants used the coca leaf from about 1500 B.C. onward. The kings, who owned all the coca plants, were considered living gods, so drug usage became mingled with religion. Coca was included in burial gifts in the graves of kings and noblemen; it was thought to give the deceased exalted well being and freedom from fatigue and hunger in the afterworld.

The Incas also used the coca leaf as an anesthetic in performing medical operations. Incan priests are known to have drilled holes into the skulls of "raptured" individuals. Skulls have been found that have holes to let out the evil spirits. From growth of bone following the "operation", it is clear that at least some of the patients actually survived the procedure.

When the conquering Spaniards learned of the power of coca leaves, they realized the economic importance of the leaf. They distributed the leaves to their enslaved workers throughout the New World in order to sedate the natives and to increase the amount of work that could be performed under very stressful and dangerous conditions in the silver and other mines. Thus, the use of the leaves spread to Peru, Ecuador, and Bolivia. This helped the Spanish in their conquest and utilization of the native peoples as slave laborers. It did not take as much food to feed these natives who were forced to work in the mines; the Spaniards fed them a few leaves with a bare minimum of foodstuffs. In addition, the populace was less likely to revolt. Today, there are many Andean peoples who chew quids of coca leaves mixed with a little alkali obtained from the ash of a wood fire. A handful of leaves is chewed daily; fresh ones are added as needed, every hour or so, to give the individual a feeling of sustained energy. Workers are able to work for hours under the most strenuous and uncomfortable conditions. It is interesting to note that many Native South Americans carried and even today carry a small rabbit-skin bag hanging around their neck containing their coca leaves.

Cocaine became popular in Europe only during the latter half of the nineteenth century after chemists were able to extract the cocaine from the coca leaves in the 1840s. This alkaloid was used extensively with even Sigmund Freud using it to (unsuccessfully) counteract the effects of morphine.

## ■ 2.6. KHAT

The peoples of the Arabian peninsula and eastern Africa, especially in the region around Somalia, Kenya, Uganda, and Ethiopia, have chewed the leaves of a shrub, *Catha edulis,* for many hundreds of years; the leaves are called *khat,* and the plant has been cultivated for centuries in what is today Saudi Arabia and Yemen. Its use was first reported in an Arabic medical text written around the year 1100, in which it was recommended that soldiers and messengers chew the leaves to suppress hunger and fatigue, much like the coca users of South America.

The leaves are chewed by the users, who seem to prefer the young, fresh leaves as being more potent. The juice is then swallowed, while the leaves, which are sometimes swallowed, are usually spit out. In Somalia and Yemen, khat use appears to be a cultural tradition with ritualistic use in groups, often in special rooms reserved just for that purpose. The main effects reported by users are increased concentration, alertness, and self-confidence. Some users report that they feel more friendly and more at ease during khat use, which seems to contradict other reports that people under its influence become aggressive and violent.

The primary active chemical in khat is cathinone, a stimulant which is similar to amphetamine, discussed later. Chewing khat dilutes the chemical with saliva, and swallowing the chemical means that digestive acids and enzymes further dilute and lead to rapid metabolic alteration of the cathinone. Under U.S. drug regulations, cathinone is a Schedule IV drug, which means it is considered less potent than marijuana, which is a Schedule I drug. However, it appears that some pushers are actively working at concentrating the chemical for direct injection or other route of administration. The effects of increased concentration and other routes of entry to the body are not known, although it might be guessed that enhanced action should be expected. A similar synthetic chemical, metcathinone, is becoming popular in some parts of the U.S., especially California.

## ■■■ 2.7. ALCOHOL

There are many references to early civilizations using alcohol. Alcohol remains one of the most addicting chemicals commonly available. It can be produced from many plants, such as maize, wheat, grapes, potatoes, etc., resulting from fermentation of carbohydrates contained in the plant to produce ethyl alcohol (ethanol) by the action of yeasts and bacteria. Since so many different raw materials may be used to make alcoholic drinks, anthropologists have noted that most societies have locally produced drinks that contain alcohol. Depending upon the raw material and fermentation process, different flavors and tastes could be imparted to the finished product, ranging from bubbly champagne to very bitter beer. Even milk from cattle or other animals may be fermented under certain conditions to produce a beer; this knowledge was discovered by many early peoples of Africa.

Beers and wines may be produced simply by the action of the fermenting organism, but stronger drinks require distillation in order to increase the alcohol content since the fermentation organisms may be killed by too high a concentration of the alcohol, which is really a waste product of their metabolism.

Worldwide, the most serious substance of abuse is alcohol, leading to thousands of deaths, accidents, and shattered dreams. According to the National Council of Alcoholism and Drug Dependence, Inc., 11.9% of the American workforce reports heavy drinking, defined as five or more drinks per occasion on 5 or more days in the previous 30 days; up to 40% of industrial fatalities and 47% of industrial injuries are related to alcohol consumption; 70% of workers in one plant reported that it was easy for them to drink at their work stations; and the cost to American manufacturers was estimated in 1988 at $32 billion in reduced productivity compared to $7.2 billion for all other abused drugs combined.

## ■■■ 2.8. MUSHROOMS AND OTHER FUNGI

Several early peoples discovered that there were fungi that contained chemical substances which could cause hallucinations; often the trance-like condition was equated to communication with the gods. A few types of mushrooms contain hallucinogenic substances; the most common and best known is the teonanacatl or sacred mushroom of Mexico, which was discovered centuries ago by the native peoples. This mushroom contains the hallucinogenic chemicals psilocybin and psilocin.

It is not clear what role, if any, these mushrooms played in religious ceremonies although use was fairly common in certain regions of today's Mexico. It is possible that these people ate the mushrooms as part of what they considered a mind-expanding exercise but without a religious overtone.

## ■ 2.9. CACTI

Peoples of the desert found that certain cacti also contain natural hallucinogens, the most common being the peyote cactus used by ancient peoples in Mexico and what is now the southwestern U.S. The mescal buttons of this cactus contain mescaline, used to induce trances which were probably a part of religious ceremonies.

## ■ 2.10. MARIJUANA

The hemp plant was used for many purposes by ancient peoples; it could be made into a very strong and long-lasting rope, which even today is one of the best ropes for many uses. However, another use was found for the hemp plant: its leaves, stems, and flowers contain chemicals which have slight hallucinogenic ability but which give an altered sense of reality to anyone eating or smoking the plant. Marijuana was found in many locales but was most commonly used for its drug ability in Mexico, hence the Spanish name commonly used. It was probably first and primarily used for pain relief but later for escape from reality. It may have had a minor religious role, but this is not certain.

It is known that the Spanish invaders of Mexico never used the marijuana to sedate or relieve pain among the natives, possibly because of the unpredictable results which can range from euphoria to violence and because it does not impart increased resistance to pain or increased endurance.

## ■ 2.11. NINETEENTH- AND TWENTIETH-CENTURY DRUGS

Various forms of opium were quite popular in Victorian England. Many noted authors and artists are known to have been opium addicts. Samuel Coleridge (1772–1834) remarked that many of his poems were composed under opium's influence. Even the fictional character Sherlock Holmes was described as partaking of opium upon occasion.

The Chinese have used opium and opiates for many thousands of years; the infamous opium dens in which opium was either smoked or just inhaled are well-known. The so-called Opium Wars of the middle and late nineteenth century were instigated when the British demanded the rights to sell opium in China; there is evidence that the British encouraged opium use among the Chinese in order both to increase sales and to render a portion of the population more docile.

In the U.S., for just one example, Edgar Allen Poe (1809–1849) was addicted to several chemicals, including opium and absinthe, a toxic drink made from wormwood; his dismissal from West Point was probably due, in part, to his opium habit. Many others from the art world as well as the business world commonly used opiates. George Washington used, near the end of his life, laudanum, a mixture of opiates in alcohol, to relieve the pain from his false teeth made of bone.

In the U.S., following the First World War, cocaine use became a fad, especially among the socially well-to-do and aspirants. This fact, together with the lawless activities related to prohibition of alcohol, helped create a mystique about illicit substance use that it was something to do if one were rich. Most of the cocaine use was by the wealthy and middle classes; the poor used alcohol. Interestingly, the gangsters of the day were more interested in pushing alcohol than other drugs, perhaps because of the larger market and ease of production.

Other chemicals, such as marijuana, were also used but were not as common as cocaine and alcohol. The Depression and the Second World War brought about a decrease in cocaine use, but not in alcohol use. The lifting of production and sales prohibitions on alcohol actually led to an increase in alcohol consumption; it appears that, at least to some extent, prohibition did decrease alcohol misuse. Unfortunately, this "noble experiment" failed and was doomed to failure because Americans have a tradition of alcohol use in most social settings; the imposition of law totally banning even beer was simply seen as too austere. This brings us up to modern times and our chemical abuse problem.

## ■ 2.12. SUMMARY

Substance abuse began as soon as humans realized that certain materials could be used to produce altered states of consciousness or relief from pain. Substance abuse can be defined as a state of intoxication produced by the repeated consumption of drugs. Substance abuse and chemical dependency can be described as addiction to any chemical such as nicotine, caffeine, street chemicals, over-the-counter medicines, household products, or prescriptions ordered by a physician. Early people found marijuana, coffee, tea, morning glory seeds, peyote, opium, coca, khat, certain mushrooms, etc. that could cause a feeling of stimulation or release from pain or of being detached from the body. Ethyl alcohol is the most commonly abused substance in the world. Most peoples of the world have found a way to ferment numerous materials into alcohol-containing drinks.

Other chemicals discovered throughout history include: quinine, cocaine; strychnine, curare, nicotine, codeine, morphine, ricinine, atropine, digitalis, reserpine, physostigmine, and caffeine. Hallucinogenic substances such as ibogaine, peyote, marijuana, and mescaline were discovered.

Modern chemists have been able to build upon these natural products to produce many wondrous drugs for the easing of pain and curing of diseases; unfortunately, some of the discoveries have been misused and substance abuse has become a major worldwide problem.

# 3 HOW THE HUMAN BODY WORKS

## 3.1. INTRODUCTION

The human body is composed of many specialized organ systems. Unlawful as well as legal chemical substances may affect one or more of these systems. To understand the effects of chemicals, it is necessary to look at how the human body works.

Some of the larger organ systems are the digestive, respiratory, circulatory, lymphatic, endocrine, nervous, and reproductive systems. Any of these can be affected by chemical substances.

It is not the purpose of this chapter to make you a biologist; only those technical terms which are necessary will be used, and every attempt will be made to keep the topics as simple as possible.

We shall briefly look at human anatomy and physiology.

## 3.2. DEFINITIONS

Simple definitions used in this chapter are:

**Anatomy** is the study of the structure of plants and animals.
**Physiology** is the study of how plants and animals function.
**Metabolism** is the total of all the chemical reactions taking place in the body.
**Exposure** refers to ingestion, injection, or inhalation of a chemical substance.

## 3.3. INTRODUCTION TO ANATOMY AND PHYSIOLOGY

Most exposures experienced by substance abusers are by means of only a few organ systems, chiefly the respiratory system, circulatory system, or digestive system, but once chemical substances enter the body, they can be moved about and affect other systems such as the nervous and endocrine systems, far from the location of entry. This chapter will cover the normal functioning and abnormal functioning under the effects of chemical substances of the major human organ systems. Each system will be discussed in turn. The next chapter will discuss the chemical substances which can affect each of these systems.

## ▬ 3.4. RESPIRATORY SYSTEM

Humans can exist for many days without food, hours without water, but only minutes without oxygen. In addition, metabolism produces carbon dioxide which can be toxic in sufficient concentrations. The respiratory system brings the needed oxygen to the body and removes the carbon dioxide.

Humans get their energy from the "burning" of food (oxidation) in their cells. In this process, oxygen combines with food to produce water and carbon dioxide. The oxygen and carbon dioxide are carried by the blood throughout the body. Any chemical substance that is inhaled, such as cocaine, enters the body by the respiratory system and is transported by the circulatory system.

Chemical substances being inhaled are carried by the air and are breathed into the lungs by the force of the **diaphragm**, a sheet of muscles located in the body cavity, which pulls on the lungs to force air in and then pushes on the lungs to force air out.

Figure 3-1 is a simple diagram of the human respiratory system.

Air entering the nose is warmed and filtered by a series of bony shelves that are covered with mucus membranes. Larger particles of the chemical substance are partially filtered out by the nasal hairs and mucus plates. Some of the chemical substance may enter the bloodstream directly through the nose since it contains many blood vessels very close to the surface. The rest of the chemical substance then passes through the **pharynx** and past the **glottis**, a flap that prevents food from the mouth from entering the respiratory system. The chemical substance then passes through the **larynx,** which contains the vocal cords, and into the largest of the respiratory passages, the **trachea**. The trachea is a large tube, about 2 centimeters (1 inch) in diameter, ringed with bands of cartilage that prevent it from collapsing. All these passages are lined with **cilia**, which are small hair-like projections that beat like small brooms to sweep some of the chemical substance particles up and out of the respiratory tract. Unfortunately, the chemical substance then is swallowed and enters the digestive system.

The trachea branches into two major passages called **bronchi** (singular: **bronchus**), one going to each lung. These branch in turn into ever-smaller passages called **bronchioles**. All these contain cartilage to prevent collapse. The very smallest bronchioles, however, do not have cartilage rings and simply end in sacs called **atria** (singular: **atrium**),

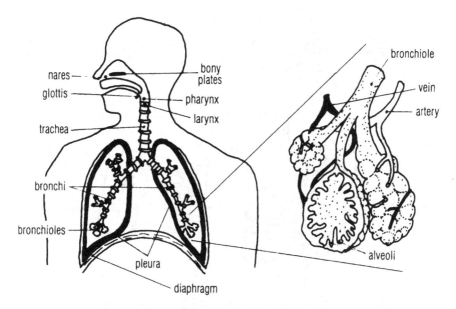

**Figure 3-1** Human respiratory system.

each of which contains many chambers called **alveoli** (singular: **alveolus**). Alveoli are shown to the side in Figure 3-1. It is at the alveoli that the exchange of oxygen to the blood's hemoglobin and carbon dioxide from the blood takes place. It is also where much of the inhaled chemical substance enters the blood stream. The alveoli on the end of the bronchiole appear like bunches of grapes on a vine. The cells of the alveoli are very sensitive to many chemicals, and the blood lies only one cell thickness away, so transfer of these substances may easily occur.

The lungs appear like pink, spongy masses in the normal person, but can become black and rigid in smokers and miners or damaged by substance abusers who inhale chemical substances. The lungs are surrounded by tissues called **pleura**, which are two sheets separated by a fluid allowing motion of the lungs as they expand and contract.

## ■ 3.5. DIGESTIVE SYSTEM

The digestive, or **gastrointestinal,** system is a long tube from the mouth to the anus. Its function is to pass food through the body and wastes out. A diagram of the digestive system is shown in Figure 3-2.

Food or chemical substances that enter the mouth pass over the **tongue** and pass the **glottis**, which acts as a flap to separate the mouth from either the digestive or respiratory systems, as necessary. The food or chemical substance is then moved by ripple-like motion down the **esophagus**, a tube leading to the stomach. Although some digestion of

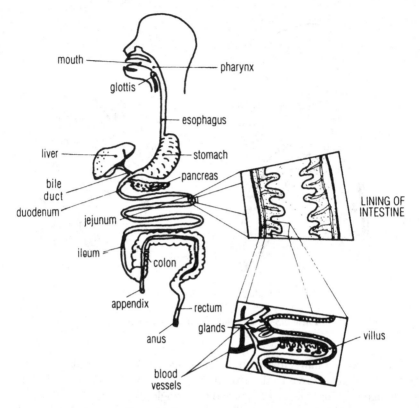

**Figure 3-2**  Human digestive system.

starch takes place in the mouth, the first real digestion occurs in the **stomach** where the food is mixed with hydrochloric acid and stomach enzymes which help break down the food. There is usually very little uptake of most chemical substances in the stomach (alcohol is an exception). The stomach is a sac-like muscle; it contacts and expands to pulverize and mix food with acid. Muscles at each end of the stomach close it off from the other parts of the digestive system.

Passing by the muscles at the end of the stomach, food or chemical substances move into the upper portion of the small intestine, called the **duodenum,** where various enzymes break down food constituents in an alkaline environment. The duodenum is the primary location where fatty foods are broken down into simpler compounds and where some substances enter the body. The duodenum is only about 25 cm (10 in.) long. The next 3-meter (10-foot) portion is called the **jejunum,** which is very rich in enzymes that further digest the food and can be the site of chemical reactions involving the chemical substance. The remaining 4 m (12 ft) of small intestine is called the **ileum**. It is the location where the food products and much of the ingested chemical substance pass to the blood through hair-like projections into the intestines. These hair-like objects are called **villi** (singular: **villus**). They are very small so they greatly increase the area of absorption.

The small intestine then joins to the large intestines, collectively called the **colon**, which are chiefly storage locations, although there is small amount of water transfer into and out of the colon. The food is mixed with mucus and bacteria and becomes waste matter called feces. When the feces reach the **rectum**, which is normally empty, the desire to eliminate becomes great (usually after about 36 hours). When the muscles of the **anus** open, the wastes are passed to the outside of the body, and the process of digestion and elimination is complete. Some of the chemical substance may pass completely through the body and be eliminated in the feces.

## ■ 3.6. NERVOUS SYSTEM

The human nervous system is very complex, and we shall only briefly discuss it. The nervous system is the target of many of the most commonly abused substances since a feeling of euphoria or peacefulness is often what the abuser seeks. The **brain** is the center of the system. It functions as a processing center, with impulses coming from receptor nerves throughout the body and going out to control all the body's functions. The brain is connected with all parts of the body by nerves; these nerves are not continuous, but have junctions where they almost, but not quite, touch. By chemical processes, electrical impulses can jump these spaces, called **synapses**. There are substances that can block the transfer of these electrical pulses. There are other substances that can cause, in effect, a short circuit. In both cases, normal nerve activity cannot take place.

The nervous system is usually divided into two parts: the **central nervous system** (often abbreviated CNS) consisting of the brain and spinal cord, which is a bundle of nerve fibers running through holes in the protective vertebrae of the backbone; and the **peripheral nervous system** consisting of nerves that connect the spinal cord with sensory receptors and glands and muscles of the body.

The central nervous system is shown in Figure 3-3. The brain has several parts, each with a different function. The chief parts are the **cerebrum**, the **cerebellum**, and the **medulla oblongata**. The cerebrum is the seat of thought, memory, and reasoning; it is the target of those chemical substances taken to remove the user from reality. The cerebellum is the seat of muscular coordination; it is the target of those chemical substances taken to relax the user. The medulla oblongata controls automatic functions of the body such as breathing, heartbeat, etc.; unfortunately, it also can be affected by many abused chemical substances, leading to respiratory collapse and even death.

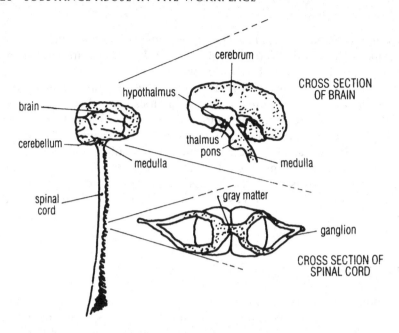

**Figure 3-3**   Human central nervous system.

## ■ 3.7. REPRODUCTIVE SYSTEM

Humans, like all higher animals, reproduce by a process called **meiosis** or sexual reproduction. All the cells of your body contain within the nucleus certain chains of chemicals called **chromosomes**. The chemical making up these carriers of genetic information is **deoxyribonucleic acid** (DNA). The ways that this chemical can arrange itself determine all the characteristics of the offspring (hair color, eye color, height, etc.). The DNA is like a template which transfers information from one generation to the next. Some chemical substances, most notably lysergic acid diethylamide (LSD), can cause damage to the DNA and the chromosomes. All the cells of the human body, except the sex cells, contain the same number of chromosomes, 46 in 23 pairs. The sex cells contain only one half this number, or 23 chromosomes. This is so because a union of the two sex cells leads to the original 46. This process is the basis of hereditary. Whether chemical substances commonly abused can cause hereditary changes (that is, effects on future generations) is a matter of speculation at this time and, while more research is needed, the possibility remains that these chemical substances could affect the heredity code and cause damage to future generations. LSD has been implicated in breaking or otherwise damaging chromosomes.

Male sex cells, called **sperm**, are formed in the male gonads, the **testes**. The process of sperm formation is called **spermatogenesis**.

These cells begin their development from precursor cells, which are formed at the edges of the testes. These precursor cells contain 46 chromosomes. They divide by a process called **mitosis** or cell division by which identical daughter cells result from a splitting of the parent cell. As the population of cells increases, the cells move to the center of the testes where they become mature sperm, following the last two stages of meiosis, and end with only 23 chromosomes. These sperm look like small tadpoles, having a rounded or arrow-shaped head and a long tail. There are chemical substances that can lead to reduction in numbers and viability of the sperm.

Female sex cells are called eggs or **ova** (singular: **ovum**) and are formed in the female gonads, the **ovaries**. The eggs are formed by a process initially quite similar to that for sperm. The immature cell divides by mitosis but, unlike the sperm, into two unequal-sized cells with one receiving almost all the cellular material. A second division of the larger cells produces another unequal division with the larger body becoming the egg and the other, like the first smaller cell, not involved in reproduction. Some chemical substances can damage the ova and make them less viable.

We now have an egg cell, which is much larger, and a sperm cell, each with 23 chromosomes carrying the hereditary coding in their sequence of genes. The union of the two cells is called **fertilization**, with the gene sequence of the new DNA decided by chance from the two parents' DNA. The details of coding and heredity are too complex to cover in this text.

The single fertilized cell then divides by mitosis repeatedly in the process of making a child. Usually this process continues in a normal fashion until birth, but sometimes there are changes because of chance, a physical cause, or a chemical cause.

The process of formation and development of an embryo is actually a programmed sequence of cell proliferation followed by differentiation (or specialization) and cell migration leading to organ formation.

The most critical period of human embryo formation is during the first three months (trimester) of pregnancy, since it is at this time that the most rapid cell proliferation takes place, upon which all subsequent development is based. The very first 2 weeks, however, is generally not a period in which chemicals can cause birth defects, since prenatal death followed by natural abortion usually occurs if damage takes place to the embryo.

A very serious matter occurs when the baby develops a chemical addiction while in the womb. Withdrawal of the chemical substance upon birth can lead to brain damage or death, since the baby is very small and unable to combat conditions that an adult might be able to tolerate, due to an immature immune system. Physical deformities are also possible if the mother abused chemicals in the very early weeks of pregnancy. Fetal alcohol syndrome and fetal marijuana addiction are today quite common; the baby usually suffers from low birth weight and slow physical and mental development if it survives. Observing a newborn shaking violently from withdrawal or seeing a dead baby is heartrending and should be more than sufficient to show prospective mothers that there are dangers to substance abuse other than to the abuser.

## ■ 3.8. ENDOCRINE SYSTEM

The endocrine system consists of those glands of the body that do not have ducts. While this might seem a strange definition, it really is not. There are many glands in the human body that produce chemicals, called **enzymes**, which promote chemical changes. Most of the glands in the body pass their enzymes to other parts of the body by ducts. For example, the salivary glands of the mouth pass saliva directly into the mouth to digest starches, and the liver passes bile to digest fats into the duodenum through a duct. Other glands, however, pass their chemicals, called **hormones**, which are often steroids, to other parts of the body through the blood. A few glands are both ducted and ductless, passing different chemicals for different functions.

There are seven ductless glands in the body, several of which can be affected by chemical substances: the **gonads**, which produce hormones to promote the development of the sex cells; the **pancreas**, which produces a hormone used in the control of the amount of blood sugar used as food by the cells; the **thyroid**, which produces a hormone used in control of metabolism; the **parathyroids**, which produce a hormone that maintains calcium and phosphorus levels; the **pituitary**, which is the most complex endocrine gland and produces growth hormones and also plays a role in control of the other endocrine glands; and the **adrenals**, which produce several hormones that function in the conversion of proteins to carbohydrates and also to stimulate the body to respond in emergencies. The approximate location of these glands is shown in Figure 3-4.

## ■ 3.9. BLOOD CIRCULATORY SYSTEM

It is believed that life began in ancient seas, and so, life even today needs to be surrounded by a mild salt solution. Every cell of the body needs to live in such an environment. Since the cells of the human body may be far removed from sources of food, there is needed some way for food materials to be brought to the cells and for wastes to be removed. The blood performs this function, but can also move chemical substances around the body.

The circulatory system is shown in Figure 3-5. The **heart** is the center of the circulatory system. It is fundamentally a pump, pulling blood from parts of the body and pumping it under pressure to other

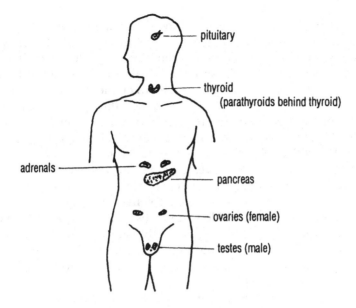

**Figure 3-4** Location of endocrine glands.

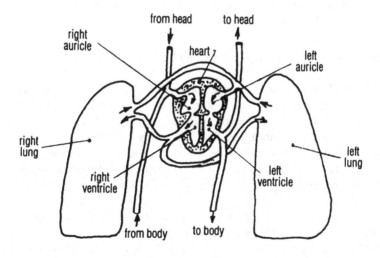

**Figure 3-5** Human circulatory system.

parts. The heart is almost totally cardiac muscle. The parts of the heart are such that they act as if all the cells were really one. Since the blood flows in a continuous circuit, we can start at any point; but, let us begin at the point in the heart where blood is pumped to the lungs.

Blood is pumped to the lungs by the right (as viewed from the front) chamber, called the right **ventricle**. The blood is pumped through the lungs where oxygen is absorbed by hemoglobin and waste gases are expelled. The now-oxygen-rich blood is pumped back to the heart's left **atrium** (sometimes also called auricle). The bicuspid value between the atrium and ventricle opens and the blood is pumped into the left ventricle. From here, it is pumped to all the parts of the body. The blood vessel first passed through is the **aorta**, the largest artery of the body. From the aorta, the blood is pumped through ever-narrowing vessels until, at last, the very tiny **capillaries** are reached. It is at the capillaries that the gaseous exchange occurs so that oxygen is given to every cell and the cell's wastes are transferred to the blood. From the heart to the capillaries, all the vessels are called **arteries**. As the blood leaves the capillaries, the return system is used. All the vessels from the capillaries back to the heart are called **veins**. Thus, the blood in the arteries is rich in oxygen and low in wastes, but the blood in the veins is low in oxygen and rich in wastes. The blood moves into ever-enlarging vessels until it returns to the right atrium through the **vena cava**, the largest vein of the body, from which it is pumped to the right ventricle. The cycle then repeats itself.

Many factors influence the rate of the heart's pumping and the pressure of the blood in the vessels. There are many chemical substances that can affect either or both; many tranquilizers and beta blockers reduce blood pressure as well as slowing heart rate. There is some evidence that marijuana also does both.

## ■ 3.10. LYMPHATIC SYSTEM

There is another system, sometimes called the body's second circulatory system: the lymphatic system. This represents another route by which fluids can flow from the cells into the blood. Larger molecules, such as proteins, cannot pass from the cells into the blood in the capillaries. However, there must be a way for them to return to the blood. This is one of the functions of this system. The fluid carried by this system is called **lymph** and is produced throughout the body, but chiefly by the liver and intestines.

Protein molecules are carried by the lymph from the cells back to the blood. The vessels of the lymphatic system are similar to those of the blood circulatory system, being larger at the points where they enter the blood stream and tiny capillaries where they contact the cells. These capillaries are different from the blood capillaries since there are minute spaces into them which can allow molecules such as proteins to enter. The lymph is returned to the blood at several locations. One-way valves allow the flow to pass into the blood but not the blood to enter the lymphatic system.

Another important role of this system is the carrying of fats and some other nutrients from the small intestines to the blood rather than by way of the villi.

Finally, very large particles such as bacteria can enter the system. As the lymph passes through the **lymph nodes**, bacteria are removed and destroyed. Sometimes, though, the lymph nodes can be temporarily overwhelmed, as during a cold. The lymph nodes can become swollen and painful. Locations of lymph nodes under the neck and armpits are commonly swollen during infections.

Effects of commonly abused chemical substances on the lymphatic system are largely unknown, although the system can carry chemicals from one part of the body to another.

## ■ 3.11. URINARY SYSTEM

Liquid wastes are removed from the body by means of the urinary system, as shown in Figure 3-6. Many chemical substances or the products of their metabolism within the body (sometimes more toxic than the chemical substances themselves) are removed from the body via the urine. In fact, the use of urinalysis for detection of substance abuse is the most common method to determine if improper chemical substances have been taken into the body. As blood passes through the **kidneys**, many waste substances are removed. These wastes, including chemical substances and chemical substance metabolic products, are carried to the **bladder**, which is a storage sac, until discharged from the body. The liquid waste is called **urine** and is generally a light yellow color from the breakdown products of protein metabolism.

The two kidneys are able to remove wastes from the blood and concentrate them for removal. The kidneys are very rich in blood vessels from which they remove the products of metabolism which would be quickly fatal if not removed. The wastes pass from the kidneys through tubes called **ureters** to the urinary bladder. Muscles at the bottom of the bladder hold the urine until it is ready to be passed. When this happens, the urine goes through the tube called the **urethra** to the outside of the body. In the female, this is a separate system; however, in the male, the urine is routed through the penis with constricting muscles blocking flow to the reproductive system.

## ■ 3.12. THE LIVER

One of the most amazing organs of the human body is the **liver**. It acts as a chemical filter to remove toxic materials from the body and then detoxifies many chemicals into less harmful substances. In addition, the liver also produces an enzyme, called **bile**, which aids in the digestion of fats.

The job of removing toxins from the blood is a very large order, especially for the chemical substance abuser. The various types of chemicals that may enter the blood stream can each require a different reac-

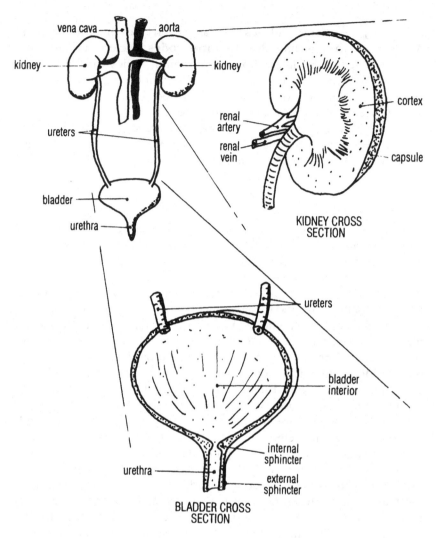

**Figure 3-6** Human urinary system.

tion to be rendered less harmful. The liver is able to detoxify many different types of chemicals; however, it can be harmed itself by some substances, either by direct action on the liver or by overwhelming the liver's capacity to detoxify. Several types of cancerous tumors can be promoted in the liver by chemicals. Long-term use of certain chemical substances, especially alcohol, can reduce the detoxifying ability of the liver or even destroy it.

In the absence of improper chemical substances or disease, the liver can usually continue its function, filtering hundreds of gallons of blood daily, and detoxifying harmful chemicals throughout a person's life. This organ is one of the most critical for life and is the primary means by which chemical substances are removed by the body.

## ■ 3.13. SUMMARY

The study of the anatomy and physiology of the human body is important if the effects of chemical substances upon people are to be understood. There are many organ systems of the body that can be affected by the effects of chemical substances, but the organ systems most commonly affected are the respiratory system, the nervous system, the circulatory system, and the digestive system.

The respiratory system functions to bring oxygen to the cells of the body and to remove gaseous waste products. The digestive system functions to bring food to the cells of the body and to remove solid wastes from the body. The nervous system acts as the communications network to coordinate the functions of the body. The purpose of the reproductive system is to produce offspring. The endocrine system functions to assist other organ systems and to control enzyme reactions in the body. The urinary system functions to remove toxic products of metabolism from the body. The lymphatic system serves to carry protein molecules back to the blood, to carry fats and other nutrients from the intestines to the blood, and to remove foreign bodies like bacteria from the body. The liver produces enzymes to help digest fats, and functions to detoxify chemicals that enter the body.

The study of anatomy and physiology will help us better understand the effects of chemical substances upon the systems of the body discussed in the next chapter on toxicology.

# 4 THE CHEMISTRY OF MEDICINAL SUBSTANCES

## ■ 4.1. INTRODUCTION

Chapter 2 described how ancient peoples, probably through trial and error, discovered that certain plants and a few animals possessed chemicals which could be used to treat diseases, reduce pain, assist in healing wounds — or kill.

Chinese writings from 2700 B.C. describe a plant known as *ma huang* being able to revive and stimulate; today, we know that this plant contains ephedrine, a powerful stimulant. Socrates died from drinking oil of the water hemlock plant as a death sentence for his teachings, which were suspected of corrupting the young. The Chinese and Koreans have eaten the root of the ginseng plant as a tonic and supposed aphrodisiac for thousands of years. Native Americans have used peyote and mescal in religious ceremonies for generations. Coca leaves have been chewed for their sedative effects, and poppy extracts have been used to relieve pain. South American tribes have used natural chemical substances to stun and kill prey for hundreds of years.

All chemical substances came either from plants, a few animals, or minerals such as metal salts or halides until the 1880s, when salicylic acid and acetanilide were first made by chemists and used as pain relievers.

All these were chemical substance uses. How do we go from chemical substance use to chemical substance abuse? Just about anything can be abused — too much food, too little food, too much sun, etc. — so it is also with chemical substances. Some people feel a need to take medications they either don't need or in amounts in excess of that required. This is chemical substance **abuse**. To better understand such abuse, it is necessary to look at the chemical and physiological actions of these substances. This chapter will discuss some of the more commonly abused chemical substances.

Use of chemical substance names is neither intended as an endorsement nor as a condemnation, but solely to alert readers to names they might be familiar with.

## ◼ 4.2. DEFINITIONS

Some of the terms used in this chapter may need defining:

**Alcohol** — technically, any of a family of organic compounds possessing a hydroxyl group attached to a carbon atom; in most common use, one of these compounds, ethyl alcohol.

**Amphetamines** — a group of compounds possessing stimulating properties.

**Barbiturates** — a group of compounds possessing sedative and relaxing properties.

**Benzodiazepines** — a group of compounds which possess tranquilizing and sleep-inducing properties and are generally safer than the barbiturates.

**Cannabinols** — a group of related compounds possessing properties of euphoria and what seems to be improved consciousness to the user; they are found in the hemp plant.

**Coca-derived chemical substances (cocates)** — a group of compounds obtained or derived from the coca plant; the most common is cocaine.

**Hallucinogen** — any chemical substance capable of producing visual or auditory effects, or space and time distortions.

**Narcotic** — a legal term for any chemical substance derived from opium or coca, as well as any synthetic chemical substance having properties similar to those of morphine.

**Opiates** — a group of compounds obtained or derived from the opium poppy plant; some of the more common opiates are opium, morphine, and heroin.

**Phenothiazines** — a group of compounds possessing properties of motor relaxation and sedation; common tranquilizers.

**Salicylates** — a group of compounds possessing analgesic and temperature-reducing properties; the most common is aspirin.

## ◼ 4.3. CLASSES OF COMMON CHEMICAL SUBSTANCES

Many thousands of chemical compounds have been developed for the benefit of people and other animals. The majority of chemical substances fall into these categories: **analgesics** to reduce pain; **anesthetics** to cause unconsciousness; **anorexics** to assist in weight loss; **antibiotics** to weaken or destroy bacteria, viruses, and other disease agents; **antipsychotics** and **antidepressants** to reduce mental problems; **anti-**

**inflammatory agents** to reduce swelling; **bronchial dilators** to improve breathing; **cardiovascular controls** to regulate heartbeat; **stimulants** to reduce fatigue; **contraceptives** to prevent pregnancy; **diuretics** to induce urinary excretion; **fertility agents** to promote pregnancy; **hormones** to assist in regulation of organ system functions; **laxatives** to induce fecal elimination; **relaxants** to ease tense muscles; **sedatives** to induce rest; **tranquilizers** to relax without central nervous system depression; and **vitamins** to assist in biochemical processes of the body.

## ■ 4.4. COMMONLY ABUSED CHEMICALS

While there are many chemical substances that possess characteristics which allow them to be abused, most fall into several categories: alcohol, marijuana, amphetamines, barbiturates, alkaloids (including the opiates), steroids, hallucinogens, phenothiazines, benzodiazepines, salicylates, bromides, and synthetics based upon one of more of the above. These will be discussed with respect to their chemical and physiological nature. Any may be unlawful or legal, even present in over-the-counter preparations.

## ■ 4.5. ALCOHOL

While the term "alcohol" actually refers to a whole class of organic chemicals, the term commonly is used for only one member: ethyl alcohol (ethanol or grain alcohol). Ethyl alcohol is a colorless liquid which mixes in all proportions in water. It is found in beers, ales, wines, and liquors such as gin, vodka, rum, rye, scotch, etc. Ethyl alcohol is probably the most dangerous substance legally available. In a 1985 study of homicides and motor vehicle accidents in Bexar County, Texas, it was found that alcohol was present in 55% of drivers killed in accidents and in 63% of the homicide victims. Deaths due to automobile accidents exceed 55,000 per year in the U.S.; it appears that alcohol is involved in over half of these deaths. In the case of homicide victims, it is also likely that the person who committed the crime was also using alcohol. Repeated studies have shown that a majority of persons arrested are positive for alcohol use.

A closely related compound, methyl alcohol (methanol or wood alcohol), is also sometimes abused but is so toxic that coma or death often result from ingestion. Methyl alcohol, also a colorless liquid, is sometimes consumed by alcoholics as a substitute for ethyl alcohol; in such cases, blindness may occur after a single small dose, with death following ingestion of between 30 and 100 grams (1 to 3 ounces). Incidentally, there is a slight concern about the artificial sweetener **aspartame** (NutraSweet®) since it can be metabolized in the body into methyl alcohol. While there are many alcohols, such as methyl alcohol or pro-pyl alcohol, when the term "alcohol" is used without further descrip-tion, it is generally understood that ethyl alcohol is meant.

Ethyl alcohol has been known for thousands of years; its name comes from the Arabic **al-koh'l** meaning "subtle". Ethyl alcohol is pro-duced by certain anaerobic (non-oxygen requiring) bacteria which use carbohydrates as their food. The carbohydrates can be of many kinds: grains such as corn, rye, barley, wheat, grapes, cherries, etc. Concen-trations of ethyl alcohol are often given as **proof** in which 200 proof refers to pure alcohol (100 percent), 100 proof to 50 percent, etc. The term "proof" originated in an early test of ethyl alcohol strength which involved soaking gun powder with alcohol and attempting to ignite it. Some people incorrectly believe that beer or wine are not "as danger-ous" as "hard liquor"; however, a 12-ounce can of beer, a 5-ounce glass of wine, and 1 1/2-ounce of liquor all contain the same amount of alcohol with the same effect on the body. Likewise, cold showers, cof-fee, raw eggs, or exercise do not quicken sobriety. Each half ounce of alcohol takes the body about 1 hour to detoxify and eliminate. Only time (and the liver and kidneys) can remove alcohol from the body.

The effect of ethyl alcohol on the central nervous system is that of a general anesthetic, not a stimulant, as is sometimes falsely thought. Symptoms of ethyl alcohol intoxication include impairment of motor skills, irrational behavior, reduction of inhibitions, drowsiness and sleep, and ultimately coma or death, although it is much less toxic than me-thyl alcohol. Ethyl alcohol requires between 250 and 500 g (10 to 20 oz.) of pure substance (200 proof) to cause death to the average adult. This means that about a quart of whiskey consumed over a brief pe-riod of time can cause death. It should be noted, however, that some alcoholics have developed a tolerance to alcohol so that more would be required to be a fatal dose. But, for young persons (due to smaller body mass) and those not habituated to alcohol, the amount might be much less than this.

Alcohol is one of the few substances which can be absorbed di-rectly from the stomach, so blood levels increase rapidly following in-gestion. Peak alcohol concentration in the blood depends on many factors (time since eating, types and quantities of foods eaten, amount ingested, etc.), but, in general, occurs between 20 and 40 minutes following the drink. The alcohol is eliminated by the body at a rate of about one half that of absorption; about 10% in the urine and 5% by the lungs with the rest oxidized, chiefly in the liver, at a rate of about one third of an ounce per hour for a normally healthy person.

Alcohol increases sensitivity to barbiturates, phenothiazines, and other central nervous system depressants. Alcohol may pass through the placenta and produce toxicity or even death in the fetus. Fetal alco-hol syndrome is often seen in newborns of mothers who have abused alcohol during pregnancy; the babies often are of low birth weight, suffer from infection, are generally weak, may possess liver or brain damage, and must suffer through withdrawal if they survive.

## ■ 4.6. AMPHETAMINES

Amphetamines are powerful central nervous system stimulants with only limited use in modern medicine (although they were formerly popular in diet medications). They are known as "uppers" in the chemical substance culture and are commonly abused, especially by persons attempting to resist sleep and drowsiness. It also appears that amphetamines may have some effect on the peripheral nervous system, leading to hypertensive crisis (large increase in blood pressure). Most amphetamines are solids and are formed into pills or capsules, often with quite colorful gelatin or coating.

Symptoms of use include restlessness, elation, intense but meaningless activity, hallucinations, increased blood pressure, and possibly panic followed by exhaustion. Pupils of the eyes become dilated (enlarged) but responsive to light. Death is possible from heart attacks.

The relationship between ingested amount and blood levels is unclear at this time, as are methods of metabolism and elimination, except it is known that about 20% is excreted unchanged in the urine (depending upon pH). Tolerance, the need to ever increase the amount consumed, is very evident in the use of amphetamines.

Evidence of amphetamine use includes possession of pills or capsules (mostly likely dosage forms), weight loss, decrease in appetite, and mood elevation.

The worker on amphetamines is nervous, often suspicious, and generally hyperactive. Amphetamine use is fairly common in long-haul truck drivers and others who have to stay awake and alert for extended periods of time. The employee abusing amphetamines may appear to be a very hard worker, even aggressive, yet is always at the brink of a collapse.

## ■ 4.7. BARBITURATES

The family of barbiturates is based on barbituric acid and are general depressants. The name comes from the fact that the first one was discovered on St. Barbara's Day. They are sometimes known as "downers", "ludes", "reds", "yellow jackets", or "tranks" in the chemical substance culture because of their effects or appearance of the capsules or pills. Depending upon the particular barbiturate, the effects can last for a very short time (Pentothal®, Brevital®, and others) to a few hours (Delvinal®, Seconal®, and others) to many hours (Veronal®, Dial®, and others). Most are solids and appear as pills or capsules with a variety of colored coverings, hence some of the common names.

Symptoms include slow and shallow breathing, smallness of pupils of the eyes, low blood pressure, appearance of intoxication such as slurred speech, increased periods of sleep, and even coma. When ingested, some of the barbiturates allow a small amount of absorption in the stomach, but most seem to be taken up from the small intestines. Some of the very short-lasting types are used intravenously, but this is fairly rare due to the brief period of effect.

Depending upon the type, elimination may be via the kidneys and urine or by destruction in the liver. They intensify the effects of other CNS depressants, including alcohol.

Evidence of barbiturate use includes taking capsules and pills (most likely dosage forms), muscle rigidity, slurred speech, uncoordinated movements, and depressed heart rate and breathing rate.

## ■ 4.8. PHENOTHIAZINES AND BENZODIAZEPINES

A variety of phenothiazine and benzodiazepine derivatives have become medically popular for treatment of hypertension and nervousness; all are sedatives or tranquilizers. Some of the compounds include: chlorpromazine (Thorazine®), chlordiazepoxide (Librium®), diazepam (Valium®), meprobamate (Miltown®), trimeprazine (Temaril®), and promethazine (Phenergan®), among others.

These compounds are typically of low toxicity with therapeutic levels much below any danger level, but toxicity can occur in young children and patients with fevers and dehydration. Deaths are very rare except as the result of accidents which occur while under the medication.

Symptoms include: hypersensitivity reactions, jaundice, and restlessness. Some compounds also produce dry mouth, constipation, and blurred vision.

The actions of barbiturates and alcohol are intensified and prolonged by the phenothiazines and the benzodiazepines. In addition, some of the minor tranquilizers (Equanil®, Miltown®, Librium®, and Valium®) have been associated with birth defects such as cleft lip; therefore, they should not be used by pregnant women, especially during the first trimester when fetal organ systems are being developed.

Evidences of use are similar to the barbiturates with slurred speech, clumsiness, confused behavior, and depressed heartbeat and respiration. Most appear as pills or capsules.

## ■ 4.9. ALKALOIDS

The chemicals known as alkaloids are synthesized by plants, and many cause powerful effects upon animals. Their name ("alkali-like") comes from the fact that most are chemically basic in nature (having a high pH value). They contain nitrogen as well as carbon, hydrogen, and oxygen. Most are insoluble in water and are bitter tasting. The role of the alkaloids in the plant is not clear although it has been suggested that they might be part of the plant defense mechanism — bitter tastes to prevent animals from eating them (although this seems unlikely, due to the addictive nature of most of the chemicals). It is more likely that they are simply metabolic waste products serving no function for the plant.

Some of the more common alkaloids and their sources are: **quinine** from the bark of the cinchona tree of South America; **cocaine** from the coca plant; **strychnine** and **curare** from strychnos plants of the West Indies and the Philippines; **nicotine** from the tobacco plant; **codeine** and **morphine** from the opium poppy; **ricinine** from the castor bean; **atropine** from the deadly nightshade; **digitalis** from the common foxglove plant; **reserpine** from the snake root plant; **physostigmine** from the calabar plant; and **caffeine** from the coffee plant. Depending upon the particular substance, many of these chemicals have been used to kill or stun prey, to relieve pain, to control erratic heart rhythms, or to seek escape from reality. There are hundreds of others, most of which possess some physiological effects.

The uses and abuses of the alkaloids are varied. Quinine was the first medicine for treatment of malaria and is still used. Cocaine has been used as a local anesthetic and for contracting blood vessels. Strychnine has found limited use as a stimulant as well as a poison. Curare and related alkaloids have been used by South American tribes to stun and kill prey. Nicotine was formerly used in several pesticide formulations but is not used today, due to its high toxicity; however, millions remain addicted to it from smoking and chewing tobacco products. Codeine and morphine have been used as sedatives and were the primary sedatives and antispasmatics of the past century; in fact, codeine is probably the most effective cough suppressant known. Some states allow the sale of codeine over the counter without prescription. Ricinine, a deadly poison, was used by the Soviet KGB in assassinations. Atropine has found uses in preventing sweating and spasms as well as an antidote in nerve agent poisoning and to dilate pupils of the eye for examination. Digitalis has been used in regulating heart rhythms. Reserpine has been used in treatment of mental illness, most commonly for hyperactive children. Physostigmine has been used by West Africans in trials by ordeal and is today used in research for nerve agent antidotes. Caffeine is used by millions of people daily as a mild stimulant; it is also a diuretic (a chemical which causes increased urination).

Many of the abusable chemical substances fall into the broad category of alkaloids or derivatives. The opiates and cocates will be discussed in greater detail in later sections because of their common appearance in chemical substance abuse situations. Most produce symptoms of euphoria, insensitivity to pain, drowsiness, nausea and vomiting, watery eyes, and a runny nose.

## ■ 4.10. OPIATES

The opiates, obtained or derived from the alkaloids of the opium poppy (*Papaver somniferum*), are among the most commonly abused chemicals. Opium, which is the raw mixture of many alkaloids, was formerly smoked or burned for inhalation, but this practice appears to be rare today. Opium has been known for at least 3000 years, but it was not until 1806 that morphine, the most active component, was isolated.

Morphine is one of the more powerful alkaloids found in opium and is commonly abused. A synthetic derivative of morphine, originally thought to be a possible means of preventing morphine addiction, is heroin. Soon, however, the addictive ability of heroin was recognized. It is one of the most dangerous of the abused chemical substances due its additive capacity; users find that they must keep increasing the amount used to obtain the same feeling of a "high". Heroin is known in the chemical substance culture as "junk", "dope", "China white", "H", or "smack", among other names. Although they are crystalline solids, morphine and heroin both melt easily and are commonly injected into the blood stream ("shooting") using a hypodermic syringe following melting in a spoon held over a source of heat. An important additional danger to the intravenous chemical substance user is the potential for transmission of blood-borne diseases such as acquired immune deficiency syndrome (AIDS) and hepatitis since intravenous chemical substance abusers commonly share needles because of the need of a prescription to obtain syringes and needles legally. Death due to accidental overdose, either from simply using too much or from a particularly potent batch, is possible; in fact, some murders have been committed by injection of heroin. Heroin is also taken by nasal inhalation ("snorting") or by injection into the skin ("skin-popping"). Some heroin is directly eliminated in the urine, while most is slowly detoxified in the liver. Recently, a mixture of heroin and fentanyl (discussed later in this section) has become popular on the streets; many deaths have resulted from its use.

Another alkaloid found in opium is codeine, which is still common in cough medicines, although there is increasing concern about its addictive ability. Codeine can be metabolically converted into morphine in the body. Most is removed by the liver with a small amount appearing in the urine. Most of the metabolic products are eliminated within about 24 hours.

Quinine, itself an opiate, is commonly used to dilute heroin; its presence in urine lasts up to twice as long (5 days) as the heroin itself and sometimes is used as an indicator of heroin abuse. Other abused opiates include methadone (used as an oral replacement for morphine and heroin but even more addictive itself), meperidine, and propoxyphene. In fact, methadone, a synthetic opiate which was developed in Germany in the 1930s and was intended to help heroin addicts break their habit, has become a legally abusable drug today with the Federal government licensing methadone clinics which dispense the drug to addicts. Unfortunately, there is little control over the program and some addicts are able to amass quantities of the drug by going to more than one clinic; these drugs either end up on the street or lead to overdoses. Deaths have been reported from overdoses of methadone.

Demerol® was synthesized as a substitute for morphine and, while it exhibits many of the same effects, is apparently less habit-forming.

Another synthetic opiate is Fentanyl® (*N*-phenyl-*N*-[1,2-phenylethyl-4-piperidinyl]-propanamide), discovered in the 1960s. Fentanyl was found to be some 50 times as powerful as morphine but does not affect the cardiovascular system as does morphine. Carfentalin® is the 4-carbomethoxy derivative of Fentanyl, and is some 27 times as powerful as Fentanyl. Other derivatives include Sufentanil® and Alfentanil®. All are habit-forming. Mixtures of heroin and Fentanyl have appeared on the street.

Symptoms of opiate toxicity include drowsiness, mood changes ranging from euphoria to depression, mental confusion, nausea, vomiting, constipation, decreased urine output, lowered body temperature, low blood pressure, and coma. The addict on opiates is unable to perform jobs which require careful attention; opiate abusers, with the possible exception of those on methadone, are usually unable to hold a job. Death, when it occurs, usually results from respiratory failure.

Effects of the opiates are enhanced by simultaneous or recent use of sedatives, chiefly phenothiazines, benzodiazepines, and barbiturates.

Evidence of opiate use depends upon the particular substance used; however, since many are injected under the skin or into veins, needle marks ("tracks") are common; also indicative of such use is presence of syringes, needles, spoons, candles or alcohol heaters, pinpoint pupils of the eyes, and cold, moist skin. The needle marks do not have to be on the arms; abusers often try to hide them by injecting into the legs or other body parts such as under the breasts, under the eyelids, on the soles of the feet, and between the toes.

## ■■■ 4.11. COCAINE AND COCATES

The alkaloids and derivatives of the coca (in the Quechuan language, *kuka*) plants (*Erythroxylon coca* or *E. truxillense*, both native to the Andes) are also varied, but the most common one is cocaine. With the exception of alcohol, cocaine is probably the most commonly abused chemical substance today. Cocaine has a very high addictive ability, with only very limited medical uses. It was formerly a component of Coca-Cola®, but was removed in the 1920s due to concern over addiction. Its systemic effect is to stimulate the central nervous system. Cocaine is usually taken as the hydrochloride salt (which is more water soluble than the pure cocaine), and sometimes mixed with baking soda. If ground up and mixed with soda, it is known as powdered cocaine (mostly abused by the middle class); if in a purer crystalline lump, it is known as "crack" cocaine (mostly abused by poorer classes). Various other salts are possible and sometimes seen; together these are called the cocates. Cocaine is a very addictive and deadly drug. It is taken orally (smoked), inhaled (snorted), or injected.

Since mid-1984, there has been a version of cocaine on the streets that is called "crack" or "rock", distilled from cocaine hydrochloride. These appear as thin slivers or small chunks resembling rocks. The pure beige crystals are usually sold packed into transparent vials resembling large vitamin capsules. Purities in excess of 90% are not uncommon. The product is usually smoked with marijuana or regular tobacco in a pipe, referred to by crack users as "free-basing", different from the free-basing in which cocaine is burned off with ether. As cocaine is one of the most powerful stimulants and because "crack" is so pure, it creates a surge throughout the user's system causing the heart to pump too fast without adequately filling up with blood, causing death very quickly and without warning.

Symptoms of cocaine or cocate intoxication include: restlessness, excitement, anxiety, talkativeness, headache, vomiting, stomach pains, and convulsions. Death, when it occurs, is very rapid, usually within 1 hour. The public press has reported about healthy sports figures who have died from a single use of cocaine, usually the crack form.

Cocaine is absorbed readily through the nasal lining and the mouth and throat; less is absorbed in the stomach and intestines since the cocaine is destroyed by conditions there. The cocaine which enters the blood is detoxified by the liver. Very little of the unmetabolized chemical substance is eliminated in the urine.

Cocaine is usually illicitly used by inhalation ("snorting") or intravenously ("shooting"). Many cultures and substance abusers smoke cocaine, often mixed with tobacco or other leafy material. Sometimes a mixture of cocaine and heroin is injected and is called a "speedball".

Evidence of cocaine use includes small glass vials, glass pipe or mirror, razor blades, syringes or needles, and needle marks.

## ▬ 4.12. HALLUCINOGENS

There are only very limited medical uses, if any, for the psychotogenic or hallucinogenic chemical substances such as ibogaine, MDMA (also known as ecstasy), peyote, mescaline, and lysergic acid diethylamide (LSD — the "S" comes from the German word for "acid", which is also a common name); MDMA was banned for human use in 1985 and LSD earlier in 1966. These substances are sometimes used in religious or pseudoreligious rites for "mind-expanding". Although persons under their influence may feel that their powers of observation, artistic abilities, or philosophical thoughts are increased, it has been shown that few, if any, of these are real. For example, a user of LSD may feel that he or she has created the most beautiful painting ever while under the influence; upon coming out, even they see that their "work of art" is not even good. Recently, however, the U.S. Food and Drug Administration (FDA) has begun to look at use of LSD and MDMA to promote self-insight and help patients deal with terminal illnesses; the outcome of this investigation is hard to predict although it is safe to say that, even if approved, usage would be highly restricted.

Symptoms include euphoria, disturbance of concept of space and time, psychosis, visual hallucinations, tremor, synaesthesia (altered sensory perceptions such as smelling colors and seeing sounds), and extroversion. The hallucinogens are usually not fatal of themselves, but can lead to accidents such as walking in front of moving vehicles or falling down stairs. Very little is known about how they function or are metabolized. LSD seems to be absorbed by and concentrated in the liver with very little appearing in the brain or brain fluids. Phenothiazines may counteract some of the effects of hallucinogens. The use of LSD was very popular in the 1960s but seems to have fallen out of common use during the next 20 years, possibly due to discoveries that hereditary material (chromosomes) could be damaged by LSD usage as well as the occurrence of "flashbacks" in which effects appear some time after usage. These facts spread quickly in the chemical substance culture. Unfortunately, the movement away from LSD may have played a role in increased popularity of phencyclidine (PCP) and MDMA, which have similar effects. Recently, however, LSD is again becoming popular, especially in California, where it is used to allow young people to remain awake and hallucinating for days at a time during extended rock concerts. LSD is commonly taken as a liquid (sometimes deeply colored blue), licked off paper decals, or placed on sugar cubes. LSD was banned in 1966 by the U.S. FDA because of its potential hazards.

PCP is a synthetic chemical more accurately named 1-(1-phenyl-cyclohexyl)piperidine. It is also known as "angel dust" or "HOG". PCP is a colorless or white crystalline solid. The hydrobromide and hydrochloride salts are medically used as depressants. The ethylamine, thiophene, and pyrrolidine analogs are apparently the more powerful hallucinogens. PCP precursor chemicals, such as piperidine, have been controlled substances since 1976 in order to reduce the availability of PCP.

MDMA or $N,\alpha$-dimethyl-1,3-benzodioxole-5-ethanamine apparently works by releasing chemicals such as serotoin from brain synapses. MDMA is an oily substance when present as the free base, but is usually used as the hydrochloride salt, a white crystalline solid sometimes called "ecstasy", which is more soluble in water and easier to consume.

In addition to hallucinogenic properties, it is also relatively toxic. MDMA use has been banned by the FDA since 1985.

Marijuana is a much weaker hallucinogen than the others mentioned in this section, and, since it is so common, will be discussed in a separate section (4.15).

Certain mushrooms also contain hallucinogenic substances. The most common is the teonanacatl or sacred mushroom of Mexico which contains two chemicals capable of causing sensory alterations; these are psilocybin (3-[2-(dimethylamino)ethyl]-1$H$-indol-4-ol dihydrogen phosphate ester) and psilocin, which is the parent compound of psilocybin and is not the dihydrogen phosphate ester. Some cacti also contain natural hallucinogens, the most common being peyote. The mescal buttons of the peyote cactus (species *Lophorphora williamsii* Lemaire) contain mescaline (3,4,5-trimethoxybenzeneethanamine), a powerful hallucinogen. The symptoms of mescaline use are virtually identical to those of LSD use. The use of such mushrooms and cacti has been a legal problem in some places, since they are used in religious practices of some Native North American tribes. Whether the government can control or ban such use is unclear although the courts have generally allowed their use in religious ceremonies.

It is clear that an abuser of hallucinogens, when under their influence or even during "flashbacks", is unable to conceive of reality and, therefore, usually is not able to perform required duties of employment.

## ■ 4.13. SALICYLATES

Although not one of the classes of drugs abused in order to escape from reality, the salicylates are used to reduce pain and are commonly overused. One of the first analgesics (pain relievers) discovered was salicylic acid. However, it is not soluble in water so control of dosage was difficult. It is today only used in dermatology. For analgesic effects, it was found that the salt, sodium acetylsalicylate, was much more soluble, yet was converted to acetylsalicylic acid in water. This chemical had good analgesic properties as well as functioning as an anti-inflammatory agent in arthritis treatment and as a fever-reducing agent. This was named aspirin and is the most used chemical substance in the world. Aspirin is the most common cause of accidental deaths of children due to poisoning.

While large amounts of blood loss are rare, there is concern over the potential for stomach lesions and bleeding from excessive intake of aspirin. Some of the problems with aspirin, such as slow dissolution and stomach irritation, may be reduced by adding buffers. There is some evidence that aspirin might reduce the possibility of heart attacks for middle-aged males without a history of heart conditions.

The most common symptoms of aspirin intoxication are stomach pain, ringing in the ears, elevated temperature, and blood acid-base imbalance. Death is extremely rare except in children but is possible even for adults, due to ingestion of massive amounts resulting in acidification of the blood, gastrointestinal bleeding, and formation of ketones in the blood. There is some concern over aspirin use being connected to Reye's syndrome in small children who have or are recovering from flu or chicken pox; such children should never be given aspirin. Some people with severe asthma are hypersensitive to aspirin. Therefore, aspirin use should be curtailed for children suspected of having flu or chicken pox, pregnant women (since labor might be prolonged and bleeding increased), and people who often have bouts with hives. Otherwise, treatment of overdoses is fairly simple and usually successful if the patient is hospitalized soon enough.

## ■ 4.14. ASPIRIN SUBSTITUTES

Because of problems with aspirin, other analgesics have been developed. Although there are many names on the commercial market, most consist of phenacetin or N-acetyl-p-aminophenol (Emprazil®, Soma®, Soprodol®, etc.), acetaminophen (Tylenol®, Datril®, etc.) or ibuprofen (Motrin®, Advil®, etc.). The latter differ in their chemical and physiological actions from aspirin and from each other.

Phenacetin (the "P" in APC tablets) exhibits some of the effects of aspirin and works with aspirin to increase these effects. N-Acetyl-p-aminophenol is a metabolic product of phenacetin and has replaced it in many uses.

Acetaminophen is about equal to aspirin in ability to relieve mild to moderate pain and reduce fever; however, it does not possess anti-inflammatory ability, so it is not recommended for relief of arthritis. Although it is thought to be less irritating to the stomach than unbuffered aspirin, there is still a possibility of some irritation to the lining of the stomach. Active alcoholics should not take acetaminophen since severe liver damage might occur. Unlike aspirin, overdoses are difficult to treat and can result in death.

Ibuprofen seems to work in the same manner as aspirin and has about the same ability to relieve pain, reduce fever, and reduce inflammation of joints. Although stomach pain is the most common complication from ibuprofen, a few cases of kidney damage have been reported among chronic users. Treatment of overdoses is similar to that for aspirin.

Another recent aspirin substitute is sodium naproyen (Aleve®), (t) 6-methoxy-$\alpha$-methyl-z-naphthalene acetate sodium. It is a sodium salt of naproxen (Naprosyn®). Like aspirin, there is concern over the potential for bleeding, either as ulcers or from thinning of the blood. There may be a synergic effect with alcohol causing dizziness, vertigo, and depression.

As with all medications, there are possible side effects from these chemical substances and caution should be exercised in using any for extended periods of time.

## ▄ 4.15. MARIJUANA (MARIHUANA)

Although the chemical substance culture insists there are no hazards and a national organization lobbies loudly (and with millions of dollars in funding) for legalization of marijuana use, this chemical substance is one of the most dangerous in common use, largely because it is thought by users to be relatively harmless. In the same study of homicides and motor vehicle accidents in Bexar County, Texas, mentioned under alcohol, it was found that 38% of homicide victims and 29% of driver deaths in motor vehicle accidents were positive for marijuana. The same study found that 38% of motorcycle drivers killed in accidents were using marijuana. The correlation of marijuana use with homicides and accidents should be shocking to those who advocate legalization of such drugs.

Marijuana consists of leaves, stems, and flowering tops of the hemp plant. It is interesting that those campaigning for marijuana legalization have recently used the tactic of talking about the beneficial uses of hemp, as for making rope, as if this somehow made marijuana usage more acceptable. Certain strains of the hemp plant, especially those without seeds, contain higher concentrations of active ingredient than others. This plant, preferred by the drug culture, is called sinsemilla and is the female plant cultured so there are no male plants present to fertilize. The species of marijuana most commonly found in the United States is *Cannabis sativa*; in Asia, *Cannabis indicus*; and, in Africa, *Cannabis africanus.* There are many slang terms for marijuana in the chemical substance culture: pot, grass, weed, herb, guaca, bang, bhang, cunjah, churrus, gunjah, hash, etc. Although there are some 421 known chemicals in marijuana, some of which are known carcinogens, the active chemical is thought to be one or more of the tetrahydrocannabinols (THC), oils which possess physiological activity, most likely the δ9 form. Extracts of the oil, called *hashish*, are particularly potent. The exact mechanism of cannabinol intoxication is not known but is apparently different from most of the other abused chemical substances.

The only legal medical application for marijuana is to reduce the effects of chemotherapy for cancer patients although some patients have been given THC to reduce the pressure in the eye from glaucoma.

Marijuana is habit-forming, although less so than heroin. Symptoms of marijuana intoxication include: confusion, apparent heightened sensations, hallucinations, ego-inflation, euphoria, dry mouth, red eyes, hunger, excessive laughing, and coma. Death, although possible, is rare, except for accidents while under the influence of the chemical substance. A panic reaction is fairly common. The ability of the chemical substance to make people feel they have greater abilities than they really do often leads to reckless behavior and accidents. Marijuana has also been implicated in causing damage to chromosomes, with the possibility of causing birth defects, a condition known as "marijuana syndrome". There is also a possibility of damage to the sex organs and the brain from long-term use of marijuana, but these are less certain.

Evidence of marijuana use includes rolling papers, pipes (sometimes containing a resinous deposit from burning marijuana), dried plant material (leaves and stems), various holders to allow smoking to the end of the cigarette ("roach clips"), and the odor of burning rope in room or on clothing.

The abuser, under the influence of marijuana, is unable to relate to reality and has clouded judgment. While the effects are usually gone in a few hours, damage to brain cells can result in impaired reasoning ability after years of marijuana use. Disruption of the concept of time and space under its influence means that the abuser should not operate equipment or drive. Other common symptoms include night blindness, altered perception of space and time, fear, withdrawal from society, talkativeness, silliness, weight loss, red or blurred eyes, over-reaction to criticism, and apathy. Lung damage is often present; there is evidence that marijuana use increases the possibility for lung cancer as well as for emphysema and bronchitis. Chronic marijuana use also decreases the number of white blood cells necessary for the body to fight off disease and infection. There is evidence that babies of chronic marijuana-using mothers exhibit symptoms similar to those of alcoholic mothers such as low birth weight, mild retardation, and weakened immune responses.

## ■ 4.16. STEROIDS

Like the salicylates and other minor pain relievers, steroids are not used to "get high" or escape from reality; however, they are commonly misused with harmful results. The term "steroid" is given to the class of compounds which includes sterols, bile acids, sex hormones, and even toad poisons, and refers to the fact that they are easily crystallized solids. They are widely distributed in plant and animal tissues. All these compounds resemble cholesterol, which is the major constituent of the spinal cord of mammals and can also accumulate as kidney or bladder stones or in blood vessels, where it can lead to atherosclerosis which has been implicated in heart attacks.

Steroids play many varied roles in metabolism. Many of the body's hormones which control growth, sexual development, etc. are steroids.

Steroids control both constructive metabolism (*anabolism*) and destructive metabolism (*catabolism*). It is the first of these roles which has led to chemical substance abuse of steroids.

Athletes desiring increased muscle mass often take anabolic steroids (those which function to build up more complex molecules) with the goal of increasing their muscles. Unfortunately, there are undesirable side effects from such use: changes in metabolic balance of the body, temporary or permanent effects on the sex organs and in secondary sexual characteristics, and possible liver damage.

Although many athletes maintain that they have greatly built up their bodies using anabolic steroids, the scientific evidence is not as clear. However, at this time, it is felt that the risks do not override the benefits, if any. Most national and international sports organizations forbid the use of steroids because of the possible dangers to the user.

However, some steroids *are* beneficial in medicine: certain adrenal cortex steroids, particularly hydrocortisone and synthetic analogs, are known as corticosteroids and are able to reduce the inflammation and pain of arthritis and skin eruptions as well as swelling in mononucleosis (glandular fever). Other steroids known as progestational steroids function as oral contraceptives although they operate by interfering with normal endocrine function, often leading to imbalances in physiology (such as mood changes, irritability, etc.).

## ■ 4.17. ANTIHISTAMINES

Commonly available cold medicines often contain antihistamines. Most cause sleepiness rather than escape from reality and are generally not considered as drugs of abuse. They are, however, often misused, with the result that employee productivity is reduced and the potential for accidents greatly increased.

Since these chemicals are known as *anti*-histamines, there must be a histamine. There is; histamine is a substance produced in the body which has a powerful effect on the blood vessels, smooth muscles, and endocrine glands. It makes blood vessels smaller, leading to an increase in blood pressure; if these vessels are in the brain, a headache results. Smooth muscles, especially the bronchioles of the lung, are stimulated, resulting in respiratory distress. It causes the endocrine glands, especially the ones which supply digestive juices to the stomach, to secrete in large amounts, resulting in stomach pain and hyperacidity. Histamine can also stimulate nerve endings, resulting in itching. Histamine is especially produced as a response to allergies. Some people experience life-threatening histamine reactions from insect stings or snake bites.

Many chemical substances have been developed to reduce or eliminate the unpleasant effects of histamine in the body. Most bear a chemical similarity to histamine, so it is thought that they interfere by blocking receptor sites on the cells. The antihistamines are based on ethylamine; some of the more common ones are diphenhydramine (Benedryl®, Sominex®, Dramamine®, etc.), chlorpheniramine (Chlor-trimeton®), triprolidine (Actidil®), and pseudoephedrine (Sudafed®). The latter two make up Actifed®. These antihistamines are beneficial when used to reduce the symptoms of hay fever or allergies; however, since histamine is not produced with the common cold, such chemical substances do not really do much to help cold symptoms other than slightly drying the nasal passages. Antihistamines are also found in some sleep aids (Nytol®, Sleep-Eze®, Unisom®, and others); however, some children and the elderly become aroused rather than calmed by these chemical substances.

Symptoms of overdose include drowsiness, dryness of the mouth and throat, and reduced sensory perception. Death is extremely rare except as a result of accidents. Persons using antihistamines should refrain from operating equipment and driving vehicles.

## ▬ 4.18. BROMIDES

The bromides are not commonly abused today, at least, not in the sense of illicit chemical substances; however, fatalities and overdoses do appear. Bromides were commonly used around the early years of the twentieth century in a myriad of proprietary mixtures; today such medications such as Bromo-Seltzer® and Sleep-eze® are less commonly used than other sedatives. You might recall that the term "bromide" has come to mean anything which is dull or boring.

Symptoms include slowed breathing, disorientation, reduced blood pressure, and coma. The response of individuals to bromides varies; some alcoholics are very sensitive to bromide intoxication.

The action of bromides is to replace chlorides in the body, producing central nervous system depression. Elimination from the body is rapid at first but then decreases; complete removal may take as long as a month. Most of the rare fatalities occur as a result of accidental ingestion by children.

## ▬ 4.19. LITHIUM

Like bromides, lithium intoxication is rare today, although there was a time when many over-the-counter preparations commonly contained lithium, with overdoses being fairly common. Today, lithium in various compounds is usually employed as a substitute for or supplement to the major tranquilizers in control of manic-depressive mental illness, most often in a hospital.

Persons exhibiting unusual behavior might have a lithium imbalance and require treatment in order to find the best balance of electrolytes, including lithium, in their blood serum.

Symptoms of lithium intoxication include: nausea, loss of appetite, diarrhea, thirst, confusion, and convulsions. Fatal doses are very rare but are possible. Effects of toxicity (and recovery) are usually rapid once the usage has ceased.

## ■■ 4.20. DIET MEDICATIONS

In the past, many diet preparations contained amphetamines; however, today the most common ingredient is phenylpropanolamine (PPA) which is both a stimulant and a decongestant. Some studies have indicated that PPA, at doses just three times that recommended by diet regimens, could produce severe or even life-threatening hypertension (increased blood pressure). Therefore, persons with high blood pressure should avoid diet medicines containing PPA (Appedrine®, Dietac®, Dexatrim®, etc.). Since PPA is also found in many cold medications (Alka-Seltzer Plus®, Contac®, Dimetapp®, etc.), persons taking diet pills with cold medicine could easily receive an overdose.

As in many other cases, there are hazards in mixing chemical substances. Many people feel that over-the-counter chemical substances are harmless; this of itself is not true, and the danger becomes even greater when different chemical substances are taken at the same time.

In general, the most effective diet is one in which the amount of food is decreased (slowly) and exercise is increased (also slowly). The use of chemical substances in dieting is a hazard since many overweight people have concurrent high blood pressure.

One fairly common medication used in dieting and often misused is **thyroxin**. One of the chemicals produced by the thyroid gland, thyroxin acts as a regulator of metabolism to increase metabolic activity. It is sometimes prescribed in treatment of obesity. Since it does tend to increase metabolism, some people feel that it can be used as a diet medicine. Use in cases other than for gross obesity is not recommended, since there are other effects resulting from increased rates of metabolism such as heart rate and blood pressure. In addition, there is evidence that thyroid medications can cause mood swings, especially resulting in nervousness and hyperactivity.

Use of any diet medication should only be done under the supervision of a physician, preferably one who specializes in dieting and weight control.

## ■■ 4.21. LAXATIVES

Most people do not consider laxatives when it comes to abused chemical substances; however, misuse of chemical substances does not have to be only for "getting high" or "mellowing". Overuse of laxatives can be quite harmful and have an effect on an employee's work performance; further, a supervisor should also consider the fact that some-times constipation is caused by chemical substances; codeine, opium, oxycodone (Percocet® and Percodan®), imipramine (Tofranil®), chlor-promazine (Thorazine®), and many others can cause abusers to also misuse laxatives.

Chronic laxative abusers may suffer from inflamed lining of the gastrointestinal tract, reduced muscular reflexes and strength of the tract, cramping, loss of potassium, hemorrhoids, and requirements for ever-increasing strength and amount of laxative.

Since there are over 700 different over-the-counter laxatives, there are many different chemicals being used. Laxatives which function by stimulating the bowels should only be used under the recommendation of a physician. Some of the more common bowel stimulants include: phenolphthalein (Ex-Lax®, Correctol®, Laxcaps®, etc.); bisacodyl (Dulcolax®, Carter's® Little Pills, etc.); senna (Gentlax®, Senokot®, etc.); cascara (Nature's Remedy®, etc.); magnesium citrate; magnesium hydroxide (milk of mag-nesia); and sodium phosphate (Fleets® Phospho-Soda, etc.). Painful cramp-ing may result from use of these as can laxative dependency. In addition, people with kidney disease should not take any of the magnesium-con-taining laxatives as kidney damage might result.

In short, there are very few medical dangers from constipation; only stool softeners (Colace®, Regutol®, etc.) and bulk-increasers (Fiberal®, Metamucil®, etc.) should be used and only when necessary.

## ■ 4.22. VITAMIN ABUSE

Like laxatives above, most people would not include vitamins in a list of abused chemical substances. True, they do not give a high or pro-duce euphoria; yet, vitamins are not without dangers.

The first vitamins discovered were all chemically amines (organic compounds which contain nitrogen); hence the early name "vitamines" since the Latin "vita" means "life". Later, others were found to be differ-ent chemically so the name was changed to vitamin. These chemicals promote chemical reactions in living cells and are needed only in very small amounts. There are some thirteen known to be needed by hu-mans. These are often divided into fat-soluble (A, D, E, and K) and water-soluble (the B vitamins and C). There are eight B vitamins: thia-min, riboflavin, niacin, $B_6$, folic acid, $B_{12}$, pantothenic acid, and biotin.

Probably the most dangerous is vitamin A, which can cause condi-tions ranging from dry skin to liver damage plus suspected adverse effects on the central nervous system. Women taking oral contracep-tives should be careful about possible vitamin A toxicity, since blood levels are increased by hormones — also, there can be possible toxic effects on the fetus. Fortunately, symptoms in adults usually disappear rapidly after stopping the intake.

In general, there is little need for additional intake of vitamins since foods supply more than needed.

## ■■ 4.23. DESIGNER CHEMICAL SUBSTANCES

With increasing frequency, illicit chemical laboratories are turning out new derivatives and new compounds. Since one aspect of chemical substance abuse is legal and laws must be very specific, a slight change in a compound can create a new chemical substance which is technically not unlawful (at least, until new laws are passed). These slight changes are known as analogs.

In addition, chemical substance addicts often build up a tolerance for a given chemical substance and seek more powerful chemical substances in order to reach whatever feeling is desired. For these reasons, and for financial ones, the creation of new chemical substances is an ongoing problem. Most of these chemical substances, being fairly complex organic compounds, are known by street names. One of these is "angel dust" or phencyclidine (PCP), a hallucinogen, which as discussed earlier, functions like LSD but apparently without the damage to chromosomes, although this possibility exists.

The majority of these so-called "designer drugs" are, however, derivatives of those above rather than totally different substances. In most cases, they function in a manner similar to their parent compound.

## ■■ 4.24. COMMERCIAL PRODUCTS

There are many commercial products which are sometimes abused. A few of the more common include: breathing of gasoline or kerosene vapors; sniffing of glue solvents; inhalation of commercial solvents such as methyl ethyl ketone or the chlorinated solvents such as trichloroethane, tetrachlorethylene, or perchloroethylene (dry cleaning solvent); breathing of ethers and lighter fluids; consumption of antifreeze and paint thinners; drinking vanilla or maple extracts for the ethanol; misuse of amyl nitrate ("poppers"), consumption of Sterno® and other methyl or ethyl alcohol-containing products; inhaling of aerosol propellants, especially spray paints (popular of late among school students), and even inhalation or ingestion of commercial pesticides. Possible interactions of chemical substances of abuse with common chemicals is the subject of the next chapter.

The list of chemicals which have been or potentially could be abused is almost endless.

Persons seeking relief or escape are capable of finding some chemical which they will try to use. Sometimes coma or death results; even with casual usage, detrimental effects often occur.

## ■ 4.25. OVER-THE-COUNTER CHEMICAL SUBSTANCES

Many of the chemical substances discussed in this chapter are available in some form over the counter without the need for a prescription; a few have been mentioned above. Many people assume that these chemical substances are therefore safe or even harmless. A pharmacologically active substance always possesses some dangers; some people are sensitized to certain agents; interactions between or among chemical substances are possible; use other than that for which the chemical substance is intended; use of excessive amounts; reduction in efficacy when really needed; and development of tolerance are all possible. Even codeine is available over the counter in many states.

Workers often abuse legal, over-the-counter chemical substances as well as alcohol and nicotine. Employers and fellow workers should be ever mindful of the possibilities if a worker appears to exhibit any of the symptoms of chemical substance abuse. The abuser needs help even if he or she may not want it.

## ■ 4.26. UNUSUAL DRUGS

Down through the ages, people have experimented with various plants and chemicals to give stimulation or relaxation. Following is a list of some of the more unusual substances that people have attempted to use:

- Mormon tea is a brewed beverage made from the Jackson weed, a plant found along migration trails across America. The main active ingredient in the plant is ephedrine.
- Jackson pepper is a bell pepper cut into sections, allowed to dry and mold. The pepper is then ground and mixed with marijuana or tobacco and smoked.
- Some young people have been known to grind up banana peels and smoke them.
- Some people steam artichoke hearts to breathe the vapors.
- The heart of a head of lettuce is sometimes dried and smoked.
- Some people chew the Kansas ditch weed that is found along roads in the Midwest in order to experience hallucinations.

## ■ 4.27. MIXING DRUGS

Unfortunately, many chemical substance abusers do not misuse just one chemical but a number of them, either at the same time or at different times. Studies at hospital emergency rooms indicate that over 10% of drivers killed in motor vehicle accidents had more than one drug in their system at the time of the crash. Some studies have indicated that between 10 and 20% of all "under the influence" arrests involve mixing other drugs with alcohol.

In most cases, the effects of mixing chemicals are not known. The discussion of antagonistic and synergistic effects was presented earlier. In some cases, the effect of the mixture is to greatly increase the effect of one or more of the chemicals; in other cases, completely different effects are produced, sometimes even death. As mentioned earlier, a mixture of heroin and Fentanyl has been responsible for hundreds of deaths. A mixture of cocaine and heroin ("speedball") is commonly injected by addicts seeking a greater response than from either alone. Some drug mixtures on the streets contain added pesticides or other poisons in order to intensify the response to the drug.

Combining alcohol with other drugs can multiply the sedative effect of the alcohol with reduced attention and vision difficulties. Mixing any drugs, including over-the-counter medications, can be dangerous, with death possible.

## ■ 4.28. SUMMARY

This brief chapter has not been intended to cover all the possible chemical substances of abuse or even all the symptoms and methods of abuse; it has been intended to present a short introduction to the chemistry of chemical substances which are most commonly abused. Members of the chemical substance culture could probably name a dozen chemicals not covered here that they could obtain on short notice; new chemical substances are being "designed" in illicit chemical laboratories all the time. New derivatives of old chemical substances and more powerful new chemical substances are introduced regularly. However, these chemical substances are usually intended to bring about a feeling of euphoria, escape from reality, relaxation, or stimulation. Any of the these symptoms, especially when they represent a change for the individual, should be reason to suspect chemical substance use. But, confrontation may often be as bad as doing nothing; try your best to convince your co-worker, worker, or employee that you really care about their health and happiness and want to help, not preach or condemn. Later chapters will assist in how you go about confronting and helping the chemical substance abuser suspected in the work place.

Additional information is available in medical, toxicology, or pharmacy texts should more in-depth data be desired, or, see your physician, who can give you valuable information on any medications you might be taking or considering. Taking a mixture or several drugs at the same time can be particularly dangerous.

# 5 TOXICOLOGY OF CHEMICAL SUBSTANCES

## ■ 5.1. INTRODUCTION

This chapter will discuss the harmful and toxic effects of commonly abused and misused chemical substances on the body of the abuser. The intent is not to make you a toxicologist, but only to give you the basic information needed to understand how chemicals can affect living systems.

## ■ 5.2. DEFINITIONS

Some of the terms in this chapter may be new to you; here are a few definitions.

**Toxicology** is the study of the effects of chemicals on living organisms.

The **threshold dose** is the minimum amount of a chemical which causes a stated response (sickness, death, soreness, etc.).

**Exposure** occurs whenever the abuser takes the chemical substance into his or her body.

**Drugs** are chemical agents which can have an effect on the body.

**Chronic** refers to longtime exposure, repeated exposures, or long periods of time between exposure and injury.

**Acute** refers to brief exposures or short time between exposure and injury.

**Teratogens** are substances which can cause birth defects; "terato-" means "monster".

**Mutagens** are substances which can cause genetic changes in future generations; "muta-" means "change".

**Carcinogens** are substances which can cause abnormal cell growth or development; "carcino-" means "cancer", from the Latin word for "crab" since such growths often look like these creatures.

## ■ 5.3. HEALTH HAZARDS

There are chemical substances which can affect living things so as to harm them. Some act slowly (chronic); others act rapidly (acute). There are many human body systems which can be affected by chemicals, but most abused chemical substance exposures occur by way of the respiratory tract, circulatory system, and digestive tract. The ways in which chemical substances can affect the organ systems of the human body are varied and often depend upon individual characteristics such as size, weight, age, etc.

In the previous chapter we learned of the various organ systems of the body; let us now consider how chemical substances can affect each of these systems.

## ■ 5.4. ACUTE EXPOSURES

The length of time that a chemical substance has to act can have a pronounced effect on its effects. In toxicology, acute effects are generally defined as those resulting from a single exposure, or multiple exposures within 24 hours or less. Sometimes, for a given chemical substance, the acute effects can be quite different from the chronic effects.

One important factor in acute exposures is the speed at which the chemical is absorbed. If a substance is rapidly absorbed, the effects are likely to occur almost immediately.

Another factor which must be considered is the frequency of exposure. Two exposures to one half the amount of a substance usually produce less effect than one larger exposure. This may be due to the body's ability to change or alter some of the chemical substance with time. Also, the body may be able to repair some of the damage if sufficient time passes between exposures.

Most chemical substance abusers seek the acute effects of the substance, whether to induce euphoria, sleep, or activity.

## ■ 5.5. CHRONIC EXPOSURES

Chronic effects from a chemical substance may include also some immediate effects in addition to the long-term effects. There can be a wide range, from immediate effects to others taking dozens of years to appear. Therefore, intermediate terms have been introduced.

Short-term exposures refer to a week or so. "Subchronic" refers to an exposure of about 3 months.

Chronic toxic effects take place whenever the agent accumulates in a biological system; that is, the absorption is greater than elimination or metabolism. Metabolic processes involve the breakdown of certain

chemicals and the formation of others. Sometimes toxic chemical substances are either destroyed or weakened by metabolism; sometimes
they are made stronger.

Chronic exposures are much more difficult to study than acute
exposures since long time periods may be involved, and different routes
may have been used in the entry of the chemical into the body.

In general, it is not chronic effects that the abuser seeks. In fact,
such conditions as "flashbacks" from the use of LSD have led to a decrease in the abuse of that substance by the community of chemical
substance abusers.

# ■■■ 5.6. TOXIC CLASSES

Toxic chemicals can be divided into several classes, generally depending upon the effects produced. These broad classes are: **irritants,
asphyxiants, hepatotoxins, nephrotoxins, neurotoxins, anesthetics,
hematopoetic toxins**, and **lung-damaging toxins**.

> **Irritants** cause irritation of organs which they contact. They may be sub-
> divided further into primary irritants, which simply cause inflammation,
> and secondary irritants, which cause inflammation plus additional ef-
> fects.
> **Asphyxiants** deprive tissues of oxygen. They, too, may be subdivided
> into simple asphyxiants, which are physiologically inert gases and only
> displace oxygen, and chemical asphyxiants, which render the body
> incapable of utilizing the oxygen in the blood.
> **Hepatotoxins** affect the liver, often rendering it incapable of detoxifying
> poisons.
> **Nephrotoxins** affect the kidneys, altering the removal of liquid wastes
> from the body.
> **Neurotoxins** affect the entire nervous system, while anesthetics are cen-
> tral nervous system depressants.
> **Hematopoetic toxins** affect the blood-forming organs.
> **Lung-damaging toxins** can damage the lungs, but are not irritants.

These classes depend upon the target organ, that is, the organ or
system directly affected by the effects of the chemical. Other ways that
toxic substances are sometimes described are by the use of the chemical (like pesticide, solvent, food/water additive, etc.) or by the source
of the chemical (like animal, plant, organic, inorganic, etc.) or by the
effects of the chemical (like carcinogen, mutagen, teratogen, etc.).

Whenever toxic classes are talked about, you must remember that
a chemical might be described by different terms depending on the
situation. For example, a chlorinated compound might be called a carcinogen, hepatotoxin, synthetic organic chemical, or degreasing solvent, depending upon what effect is being considered.

TABLE 5-1

**POISONING POTENTIAL**

| Term | Oral Dose (mg/kg) | Median Amount for Average Human |
|------|-------------------|--------------------------------|
| Practically nontoxic | >15,000 | 2.3 pounds |
| Slightly toxic | 5,000–15,000 | 1.5 pounds |
| Moderately toxic | 500–5,000 | 7 ounces |
| Toxic | 5–500 | 2/3 ounce |
| Very toxic | 0.05–5 | 2 drops |
| Extremely toxic | 0.05–0.5 | 2/10 drop |
| Supertoxic | ≤0.05 | Trace |

## ■ 5.7. POISONING POTENTIAL

Often, in addition, toxic chemical substances are divided into classes according to their poisoning potential, or how much it takes to cause harm.

Most toxicologists have agreed to certain terms which relate to ranges of quantity of substances. These are usually based upon lethal dose, usually taken orally, for humans. It is important to remember that these are for lethal doses which are usually much smaller than the amounts taken by chemical substance abusers. However, many deaths do occur each year from lethal overdoses. The usual values are for milligrams of toxin per kilogram of body weight of the exposed individual. This is often abbreviated "mg/kg".

In this system, shown in Table 5-1, practically nontoxic substances would require that more than 15,000 mg (15 grams or about one-half ounce) per kilogram (2.2 pounds) of weight be taken into the body to cause death. Since most adult humans weigh about 150 lbs (approximately 70 kg), one would have to ingest about 2 1/2 lbs of such a substance to cause death. The next group are those considered slightly toxic, requiring 5,000 to 15,000 mg/kg. Next are those considered moderately toxic, requiring 500 mg/kg to 5000 mg/kg. Toxic substances require between 5 and 500 mg/kg. Very toxic chemicals need only 0.5 to 5 mg/kg to kill.

The next group are those considered extremely toxic, requiring only 0.05 to 0.5 mg/kg.

Chemicals which can cause death at amounts less than 0.05 mg/kg are called supertoxic. These would require less than 5 drops of liquid to kill the average human adult.

Bear in mind that these are broad classes and not exact numbers. However, these groups can be used to give an indication of the poisoning potential for comparison of two or more chemical substances.

# ■ 5.8. DOSE RESPONSE

We all know someone who seems to be able to drink alcohol all day and night without seeming to become very drunk. On the other hand, we probably know people who get tipsy after one small drink. What's going on here?

Over the years, toxicologists have come to the correct conclusion that was first voiced in the early 1500s by Paracelsus that "the dose makes the poison". Almost any substance can cause injury or death in sufficient amount.

Therefore, it is important that we consider response to different doses of chemicals, not just the nature of the chemical.

Since **dose** refers to an amount, we must first consider how we measure amounts. One of the most valuable ways to consider the amount of chemical substance needed to cause a certain response (such as death, burns, euphoria, sleep, irritation, etc.) is by the number of milligrams (or ounces or whatever) to the number of kilograms (or pounds) of body weight of the person exposed to the chemical. This gives the poisoning potential. However, this is not an individual characteristic — it is based on thousands of trials and represents an average value. An individual might differ greatly from this statistical value.

**Response** is whatever effect is being looked for. It can range from slight, temporary impairment to death; however, the specific response must be defined in order to make sense of dose-response relationships.

Most values for occupational exposures, or indeed any exposures, have been arrived at on the basis of animal studies.

# ■ 5.9. ANIMAL MODELS

Experimental results for many chemical substance exposures have been based upon models and adjusted for human characteristics. Use of human subjects in the laboratory is generally considered unethical (at least as far as causing death is concerned), so most toxicity testing uses animals. There are problems with this approach. Aside from the concerns of animal rights groups, it has been demonstrated that different species respond differently to the same chemical. Even closely related animals such as rats and mice often exhibit widely different effects from some chemical substances. There is no animal (or other) model which exactly duplicates human response. The study of abusers can be helpful, but there are problems with this approach, since purity of the chemical substance and exact quantity consumed are usually not known.

Today, animals continue to be used for testing, but with greater concern for their well-being and comfort, since there are no better models available. New techniques, for example using bacteria or other plants, are being investigated. One of the standard tests for carcinogens uses bacteria. Research toxicologists would delight in finding nonanimal models which duplicate human response if just for the fact of not having to deal with housing and feeding the animals. We shall probably see much more research in this area of substitute models in the coming years.

As a dose is slowly increased from zero to a maximum, the response does not appear suddenly, but is graded. For animal testing, the end point most commonly chosen is death. A dose is selected, and the response, as defined, is observed. Other doses, more and less, will be given until a range is found. This range is bounded by the dose which causes the death of all the animals to the dose at which all the animals survive. The data are usually presented as a response curve relating the amount of the dose to the percentage of the animals responding.

## ▄▄ 5.10. RESPONSE CURVE

The graph of dose vs. response is generally an "S"-shaped curve. At the upper end, it approaches but never quite reaches 100%. At the lower end, it approaches but never quite reaches 0%. It is at the low dose end that most of the experimental problems result. Responses are usually very small and might be missed. For most chemical substances it is not known if the curve really goes to zero or not. Likewise, for many chemical substances, we do not know if there exists a threshold below which very small doses do not lead to a response. Since we are unsure, we assume that there is a linear (straight-line) approach to zero and there is no threshold; that is, any dose will cause some response. A typical response curve is shown in Figure 5-1. The dose which causes the death of one-half the animals is called the $LD_{50}$, lethal dose 50%.

## ▄▄ 5.11. DISTRIBUTION CURVE

Now that we have data from toxicity testing using the dose-response curve, we can extend the results in terms of a given population exposed to that chemical substance. Individual members of a group will respond differently to the same dose; these differences are referred to as biologic variances or individual susceptibility. Some of the differences can be caused by age, sex, heredity, physical condition, diet, previous abuse, etc.

A graph of the reaction against the number of individuals responding is called a distribution graph and is usually bell shaped. An example is given in Figure 5-2. A few individuals will respond strongly and a few hardly at all. These are the tapering ends of the curve. Most individuals fall somewhere between the extremes. The members of the groups will thus show a range of responses from slight to great. We consider that the middle of the curve represents the "average" member of the group.

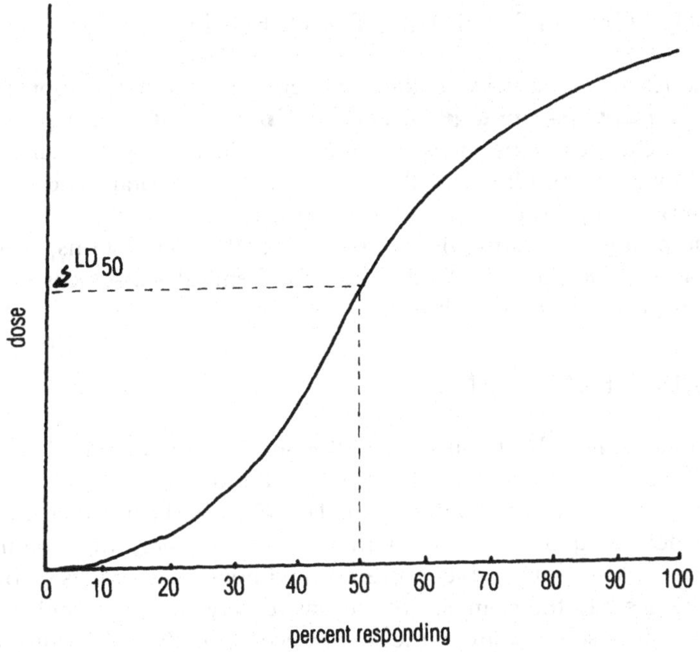

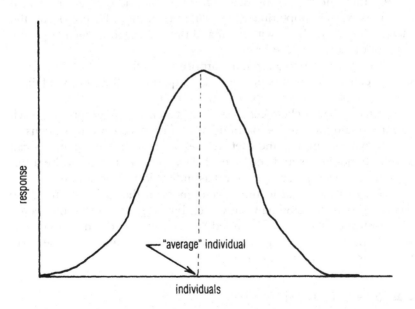

**Figure 5-1**   Typical dose-response curve.

**Figure 5-2**   Typical distribution curve.

## 5.12. CANCER AND BIRTH DEFECTS

Chronic toxins are usually divided into groups depending upon their effects: **teratogens, mutagens, and carcinogens**. It is also possible for some chemical substances to affect reproduction by reducing sex drive; however, the effects of these are usually temporary and conditions improve upon removal of the exposure.

The next few sections will discuss each of these conditions in more detail and discuss a few of the common chemical substances which can cause each of the conditions.

## 5.13. TERATOGENS

Many of you may be familiar with the sedative thalidomide, which, although never approved by the U.S. Food and Drug Administration, was used in several nations during the late 1950s and early 1960s. Over 10,000 deformed babies were born to mothers taking this chemical substance. Most of the babies were born without arms or legs or both. Some babies had their limbs, but they were very short in length.

You may also be aware of the occurrence of hundreds of malformed babies in Japan as a result of their mothers eating fish contaminated by organic mercury compounds from industrial contamination of a coastal bay. The children suffer from cerebral palsy as a result.

Another chemical substance which was formerly used to induce abortions was aminopterin. If the planned abortion did not occur, the infant was usually born with skull and brain damage, defective palate, and abnormalities of the limbs.

One proven teratogen might surprise you: ethyl alcohol. This common chemical found in beer, wine, and liquor can cause growth failure and impaired brain development in the baby.

Many common chemical substances taken during pregnancy, such as aspirin, amphetamines, barbiturates, and tobacco, are teratogens.

Heroin, morphine, and methadone can also be teratogens, as can several tranquilizers and sedatives. LSD was once thought to be a teratogen; however, recent research has indicated it may not be.

The greatest threat to the development of the baby is during the early weeks of development; however, birth defects caused by chemical substances should not be ruled out at any time during pregnancy. In fact, there are some chemical substances which can also cause abortion, even well into the pregnancy.

## 5.14. MUTAGENS

You may recall from the previous chapter that the basis of heredity is the genetic code carried by the gene sequence of the chromosomes.

There are chemical substances which can alter this code, producing mutations. Teratogens only produce their change in the next generation, but mutagens can continue to cause defects for untold numbers of generations.

Human genetic alterations may be caused by several ways in which the genetic code fails to be transmitted accurately. Chromosomes are composed of genes made of the chemical DNA located in a particular sequence. If the sequence is changed, for example, by a part being left out or extra parts being put in, the "message" will be altered and a mutation may result.

It should be mentioned that not all changes in the DNA will automatically lead to a mutation or, if they do, that the change will necessarily be harmful. However, in general, we must consider that unnecessary chemical substance exposures are abnormal and therefore the results may be harmful.

Another way in which mutations may be caused is due to breaking of a chromosome resulting in rearrangement of the pieces or even just fragments of the original. A third way that mutations are caused is through the gain or loss of complete chromosomes.

Any of these three methods can be caused by a chemical substance acting upon the sex cells of either parent; however, such changes can also take place naturally or as a result of exposure to other factors such as ionizing radiation.

One scientific study has indicated that about 0.6% of all newborn babies have some detectable chromosomal abnormality and some 0.3% vary in the total number of chromosomes. However, among spontaneously aborted fetuses, the rates are as much as 30% with defects in chromosomes. It is believed that this is one reason for such abortions.

Some of the chemical substances thought to cause mutations include caffeine, LSD, and marijuana. Most mutagens appear to cause their damage by chemical reactions with the amino acids of the DNA.

## ■ 5.15. CARCINOGENS

The preceding conditions affect future generations; however, there are chemicals which can harm the exposed individual. Some of these chemicals can cause changes in cell growth or cell metabolism. These substances are called carcinogens.

Remember, though, that there can be a wide spectrum of harm caused by these substances. In other words, "cancer" is not one disease like polio or smallpox but is many different diseases all grouped together since they all involve changes to cells. A chemical substance might "cause" one kind of cancer but not any others.

How chemical substances can lead to cancers is not clearly understood; however, it is thought that at least some cancers result from chemical changes to the DNA of cells of the exposed individual. Since these cells are not necessarily involved with reproduction, genetic changes may not occur. Other chemicals appear to weaken the body to allow other things to cause the cancer.

Cancers have apparently been with us since the beginning of time — ancient Egyptian mummies have been found with tumors. Whether these were caused by exposure to chemicals can never be known, although it is highly unlikely considering the small number of chemical substances available in those days. However, in the late 18th century, it was noted that chimney sweeps in England often suffered from an otherwise rare form of scrotal cancer, caused by years of exposure to soot rubbed into the skin.

Shortly thereafter, it was discovered that coal tar could cause cancer. Recall that coal was the primary fuel in those days, so millions of people were being exposed. Later research found that a group of chemicals called polyaromatic hydrocarbons (PAHs for short) were to blame. In the 1930s, it was observed that cancers of the urinary tract were common among German workers in the dye industry. Aromatic amines were found to be the cause. About the same time, other dyes called azo (nitrogen-containing) compounds were found which could both directly cause cancer and also predispose an individual to cancer.

Evidence of carcinogenic ability is unclear for most of the chemical substances commonly abused. It make take many years before there are enough scientific data to be sure one way or the other. However, because these chemical substances are taken for their effects on the human body, it is very likely that one or more will be shown to cause cancer in the future.

One chemical substance abuse-related situation which needs study is that of soldiers exposed to the defoliant Agent Orange in Vietnam. Many veterans of the war in Southeast Asia have claimed a variety of illnesses which have been suspected, but never proven, of being related to the chemical dioxin (TCDD), which was a minor impurity in Agent Orange. However, there have been no studies (and none are planned at this time) to investigate the possibility that these illnesses could be due to abuse of chemical substances, especially heroin and marijuana, which were commonly used in Southeast Asia, or to a synergistic effect between dioxin and these abused chemical substances.

## ■ 5.16. MIXTURES OF SUBSTANCES

Another area of concern is that there are little scientific data on the effects of mixtures of chemicals. Almost all the toxicology research conducted to date has focused on pure compounds in order to determine the effects of a particular chemical on biological systems. Unfortunately, most chemical exposures are multiple: chemicals mixed in the body from legal prescriptions, alcohol, illicit drugs, environmental chemicals, occupational chemicals, etc., about which little, if anything, is known. Sometimes two (or more) chemicals combine to make the effects of either more pronounced; this is known as a synergistic effect. In other cases, however, one chemical tends to counter the other, producing reduced effects; this is known as an antagonistic effect. There are even examples of a mixture of chemicals causing effects unrelated to those of any of the components.

Impurities in street drugs make this question of the effects of mixtures even more important. Just as moonshiners used (and still use) dead rats, sawdust, lead-soldered copper pipes, etc. with toxic results, today's drug pushers have been known to mix just about anything in the drugs they sell. Rat poison and strychnine are popular for mixing with heroin and cocaine, since they cause a physiological response that might be confused with a "high". Street drugs can contain just about anything from sodium bicarbonate to deadly poisons.

The effects of mixing chemicals in the body are so largely unknown that this could, indeed, become the largest area of toxicology research in the near future. This book touches on the possibility of effects due to environmental and occupational chemicals while under the influence of illicit or misused drugs.

## ■ 5.17. SUMMARY

Just as a basic knowledge of anatomy and physiology is important in learning how to deal with abused chemical substances, so is a basic knowledge of toxicology.

There are chemical substances which can cause harm quickly (acute) or slowly (chronic). Among the chronic toxins are mutagens, teratogens, and carcinogens.

Toxicologists use experimentation to arrive at dose-response relationships. From these data, values are determined which can be used to estimate the effects at differing levels of human exposure. Unfortunately, most the data relate to pure substances, with little known about mixtures or impurities.

# 6 INTERACTIONS WITH OCCUPATIONAL OR ENVIRONMENTAL CHEMICALS

## 6.1. INTRODUCTION

This short chapter is intended to warn the drug abuser or potential drug abuser of hazards which might not ordinarily be considered. There is an important area of study which has not been discussed either by the public or by scientists as much as it should be: many occupations involve contact with other chemicals which can be more hazardous when exposed to while taking chemical substances. The importance of occupational exposures to carcinogens and other harmful chemicals is shown by the Occupational Safety and Health Administration's (OSHA) regulations in Volume 29 of the Code of Federal Regulations (CFR) Part 1910. These regulations place strict limits on exposures to potentially harmful chemicals in the workplace, often limiting the exposure to no more than a few parts per billion or even lower. Not considered is the fact that there are chemical substances which might increase the dangers of exposure to industrial chemicals many times. In addition, these regulations are based upon healthy, nonimpaired workers. This chapter will briefly warn of the hazards in using chemical substances in workplaces where other chemicals are being used.

In addition, it is now recognized that people may have exposures, even though usually small, to environmental chemicals as a result of industrial activity, decay of radioactive materials in the soil, outgassing of solvents from paints and carpets, photochemical reactions in the atmosphere, etc. The U.S. Environmental Protection Agency (EPA) issues regulations limiting the exposure of the public to chemicals in the environment in Volume 40 of the Code of Federal Regulations, sometimes to levels of a few parts per million or even less. It is recognized that there are environmental chemicals that can be harmful above certain concentrations. There are people who are more sensitive to these chemicals than others and may suffer from watery eyes to upper respiratory distress to even more serious chronic conditions such as emphysema or depressed immune response. The combination of chemical substances and occupational or environmental chemicals can increase the danger from substance misuse.

## ■ 6.2. SYNERGISTIC AND ANTAGONISTIC EFFECTS

There are cases in which the mixing of two or more chemicals can produce increased toxicity greater than that of any of the components. This is referred to as synergy or potentiation. Workers who deal with chemicals in their routine employment are at greater risk that the drug they might be using can be increased in toxicity by the chemicals they work with. Use of drugs with occupational exposures can often lead to unpredictable results.

In some cases, the opposite may occur — a reduction in toxicity when two or more chemicals are mixed; this is known as antagonism.

## ■ 6.3. OCCUPATIONAL EXPOSURES AND SUBSTANCE ABUSE

An excellent reference on the interactions between alcohol and various industrial and workplace chemical substances can be found in Edward J. Calabrese's *Alcohol Interactions with Drugs and Chemicals,* cited in the references. Ethanol or ethyl alcohol is the most commonly abused chemical substance; for this reason, most of the scientific studies involving interactions with industrial chemicals have dealt with this alcohol. For example, it has been shown by numerous studies that occupational exposure to many metals, including cadmium (formerly quite common in welding and braising), cobalt (used in glassware and dyes, and even formerly found in some beers), lead (commonly found in plumbing, radiator work, battery repair, leaded gasoline fumes, and metal smelters), and mercury (from ore smelting, broken thermometers and other gauges, and dentistry), can increase the toxicity of ethanol, with resulting serious conditions and even death possible.

Hydrogen sulfide is commonly found in many occupational settings; concurrent consumption of alcohol can lead to more sudden and deeper unconsciousness than from either alone.

Although benzene is not as commonly used today as in the past, there are still many workplaces where benzene may be encountered; ingestion of ethanol together with benzene exposure can increase damage to the liver.

While benzo(*a*)pyrene is not usually found in industry since it is a potent carcinogen, it is possible that some workers, including researchers, may be exposed to it; a synergistic effect has been noted with ethanol.

Carbon disulfide is widely used as a solvent in industry and is known to be a neurotoxin; methanol exposure appears to increase its toxicity, while ingestion of ethanol has a much smaller, if it exists at all, increase in toxicity. However, potentiation of the effects of carbon tetrachloride has been clearly demonstrated. Caution is advised in consuming alcohol whenever exposure to carbon disulfide or carbon tetrachloride is possible.

There may also be a relationship between ethanol and carbon monoxide exposure; this is important since there are many sources of carbon monoxide in the workplace, ranging from automobile garages to sources of combustion.

Dimethyl sulfoxide (DMSO) is used in medical treatments (and sometimes as a nonapproved vehicle for transport of drugs directly through the skin). The individual taking DMSO experiences a garlic-like taste. Research has indicated that ethanol and DMSO interact in a variety of ways; interestingly enough, DMSO seems to speed up the elimination of alcohol from the body while, at the same time, increasing the effects of intoxication.

It was discovered during the Second World War that workers at munitions plants who used alcohol were more susceptible to chemical intoxication from exposure to dinitrotoluene (employed in making TNT).

While the above cases have discussed situations in which the adverse effects have been made more pronounced, there are cases in which mixing substances can actually result in reduced toxicity. The most common antifreeze is ethylene glycol, a potent kidney poison. Ethanol has been used successfully to reduce the effects of ethylene glycol intoxication. In fact, ethanol seems to reduce the toxic effects of some other compounds like monofluoroethanol and monochloroethanol and, possibly, some aromatic hydrocarbons like toluene and xylene. This does not, however, even suggest that such treatment be considered unless under the care of a physician.

Aside from acute effects, there have been many studies of the effects of ethanol in cancer production. It appears that routine ethanol use can increase the potential for cancer due to exposure to nitroamines, certain pesticides, and styrene. Exposure to vinyl chloride (used in making plastics) together with alcohol use has been shown to increase the chance of liver cancer.

Due to possible interactions among drugs, including alcohol, and occupational exposures to chemicals, care should be exercised by anyone who routinely deals with chemicals. The industrial exposure standards for chemical exposures are based upon healthy, nonimpaired workers. It is possible that two people, both of whom abuse drugs, could find one suffering much more severe symptoms or even death due to occupational chemical exposure than the other.

Although most of the studies have involved alcohol, similar interactions are possible with a wide variety of abused drugs and workplace chemicals. All workers with chemicals should exercise caution in using any drug. A prudent path might be to check with the company physician or other health care provider if you feel there is a possibility of adverse interactions between drugs you are taking and the chemicals in your workplace.

## ■ 6.4. ENVIRONMENTAL EXPOSURES AND SUBSTANCE ABUSE

As with occupational exposures, there are many questions about the possible interactions of drugs and other chemicals in the environment. There is today almost no scientific research involving the possible interactions of drugs and substances found in the environment. However, most chemical substances found in the environment are present in extremely low concentrations, on the order of parts per billion, so the risk is probably quite low.

Any drug which reduces the effectiveness of the immune system could increase the potential for harm from chemicals in the environment. Alcohol and marijuana are very likely candidates.

While there is little evidence at this time, this section is designed to remind the potential drug abuser that there are many other effects possible than simply getting high or mellow. We may find that many of the illicit drugs, due to impurities, increase the effects of environmental chemicals and might even themselves be carcinogens.

## ■ 6.5. SUMMARY

OSHA regulations place strict limits on exposures to potentially harmful chemicals in the workplace, but these regulations are based upon healthy, nonimpaired workers. Not considered is the fact that there are drugs which might increase the harm from exposures to industrial chemicals.

It is recognized that people may have exposures to environmental chemicals as a result of industrial activity, decay of radioactive materials in the soil, outgassing of solvents from paints and carpets, photochemical reactions in the atmosphere, etc. The EPA issues regulations limiting the exposure of the public to environmental chemicals which can be harmful above certain concentrations. The combination of chemical substances and occupational or environmental chemicals can increase the danger from substance misuse.

There are cases in which the mixing of chemicals can produce increased toxicity greater than that of any of the components. Workers who deal with chemicals in their routine employment are at greater risk in that the drug they might be using can be increased in toxicity by occupational chemicals. Use of drugs with occupational exposures can often lead to unpredictable results.

It has been shown by numerous studies that occupational exposure to many metals can increase the toxicity of ethanol.

Hydrogen sulfide, benzene, benzo(*a*)pyrene, carbon tetrachloride, carbon monoxide, DMSO, and dinitrotoluene, among other chemicals, have been shown to be increased in toxicity when combined with alcohol ingestion.

It appears that ethanol use can increase cancer potential when combined with exposure to nitroamines, certain pesticides, vinyl chloride, and styrene.

Care should be exercised by anyone taking drugs who routinely deals with chemicals. Most studies have involved alcohol, but similar interactions are possible with a wide variety of abused drugs and workplace chemicals.

There are also many questions about the possible interactions of drugs with chemicals in the environment. Drugs which reduce the effectiveness of the immune system could increase the potential for harm from chemicals in the environment.

# 7 IDENTIFICATION OF THE SUBSTANCE ABUSER

## ■ 7.1 WHY DO PEOPLE MISUSE CHEMICAL SUBSTANCES?

This question has plagued behavioral psychologists for a long time. Is substance abuse simply the result of peer pressure or is it a maladapted response to stress? Both of these may be valid reasons, but are they the only ones? Perhaps some people get started because they like the feeling produced even if they are not under excessive stress. Often the beginning of drug abuse is seeking the apparent pleasurable effects such as euphoria, stimulation, sedation, or hallucinations. This stage is commonly followed by physical dependence produced by prolonged use of the chemical substance whose pharmacological action causes the body to adapt to its presence (tolerance). However, the most common beginning is usually due to peer pressure or as a response to real or imagined stress. Let us first look at some of the reasons and then focus on peer pressure and response to stress.

**Peer pressure** — One of the greatest problems for young people is how to respond to the pressure to be like the kids around them. These outside influences want them to partake in the use of chemical substances. During the impressionable years, generally from 7 to 14 years of age, kids are bombarded daily by their peers to take chemical substances. On the other hand, during the last few years there have been programs such as Drug Awareness and Resistance Education (DARE) developed to attempt to reduce such influence. This and other programs have been providing training and education in how to resist negative peer pressure about chemical substances.

**Maladapted response to stress** —Another factor in why people seek escape in chemical substances is the feeling of release from ordinary standards of conduct or from stressful situations. When one is under stress at work, at home, in financial matters, or even from environmental stressors, a great deal of pressure may be exerted on an individual. Most people can identify the stressors affecting them and then adapt to the situation; but there are many people who can not adapt and will look for means to escape from the reality of the issues around them.

**Rebellion** — Some people, especially young people, feel that they have to rebel against parental or societal values or authority in order to assert their own growing up. The fact that these substances are forbidden by parent, laws, etc. tends to make them somehow mysterious and therefore wanted. This is likewise seen in many young people's choices of actors, singers, etc., some of whom are chosen to aggravate parents and other adults. Using drugs they know anger their parents makes the drug usage more fun. Even in people not so young, the allure of doing something that society frowns on makes it a challenge. Some workers misuse chemical substances, at least in part or at first, as a rebellion against the authority of their boss, foreman, etc.

**Adult pleasures** — Almost as the opposite of rebellion against parents and other adult authority figures, some young people begin their experience with drugs following the example of adults. When young people see the misuse of alcohol, tobacco, and tranquilizers by parents, they often feel that they, too, are "entitled" to the apparent pleasures of these substances. In addition, since the drug use is banned for them, but used by adults, it seems to make the drugs more alluring as an "adult" thing to do.

**Curiosity** — Some people get started with drugs simply out of curiosity about the effects of these chemicals. They hear that marijuana, cocaine, or whatever is used by thousands of people, maybe even by people they know; they wonder what the attraction is for these chemicals and then try them. In some cases, one use of cocaine has produced death; in thousands of other cases, the person becomes addicted to the drug.

**Infrequent users** — Not all drug abusers are constantly "on a high" or even under a drug's influence. Some users only use drugs on special occasions, such as parties, to stay awake studying for exams or doing homework, or to reduce pain from a headache. Others only use drugs every so often, on a spree or on vacation. Unfortunately, some people cannot limit their drug usage and develop a need to continue to use drugs, usually in ever-increasing amounts or strengths. They become addicted, either psychologically or physiologically, to the drug and develop a fear of withdrawal or absence of the effects of the drug.

## ■ 7.2. PEER PRESSURE

The dictionary defines a "peer" as "a person who has equal standing with another, as in rank, class, or age." Pressure is defined as "the act of pressing; the condition of being pressed; to force, as by overpowering influence or persuasion." With these two definitions in mind and the fact that it is human nature to want to please others, we can begin to understand the pressure that is being put on many people, especially youngsters who suffer from self doubts, today. We want to be liked by others, respected, and accepted. These are significant influences in our society. We may not admit to them, but they are there! This desire may not be necessarily a bad or good thing, but it is part of our basic need to belong. The management philosopher Herbert Maslow discussed this desire to belong in his theories of behavior. In his theory of Hierarchy of Needs, it doesn't really matter whether we joined because we wanted to or were made to; the basic fact is we want to associate with others, and this will influence our thinking and behavior. We try to accommodate our behavior to the group norm, whether this be wearing a particular style of clothes or haircut or acting in a certain way. Such needs can outweigh even basic needs of food and shelter. Accommodation to the group norm has been one of the most civilizing forces in human development; however, this force has also led to great personal conflict, loss of one's self-esteem, and, in some cases, denial of one's self worth. Look around — you see how people will alter their speech, clothing, thinking, and behavior to "fit in" with the group. No matter where you came from, your age, or your nationality, each of us tends to modify our actions in an effort to get along with the people with whom we associate, and we want to be accepted by them. This "peer pressure" is seen in a cross-section of society; it is not only found in the young but even in adult groups. The teen years and early twenties are a difficult period for most people; they are attempting to learn who they are and simultaneously desire the freedom to make choices, yet want to be accepted by their peer group. Parents are adults and not of the peer group, so their wishes and plans do not have meaning. So, while parents insist that their youth take on adult responsibilities, young people turn to their peer groups for support which they feel cannot be obtained from parents. This is a normal sign of independence, that they can function on their own and do not need their family. We have a tendency to try to shepherd our children into adulthood so they do not make the mistakes we did or the mistakes we perceive we made. Young adults do not or are not willing to tolerate this type of interference. They want to make their own decisions and lead their own lives. This desire to leave the nest can be interpreted as rebellion, "I failed as a parent," or "they do not like me"; or it can be that they need to learn how to make it on their own.

As an employer, you may see this in your younger workers. You should place your demands, requests, and rules on your employees in such a manner that you do not give the impression that you are testing their capabilities or are being arbitrary but are doing so because you are concerned for their well-being and safety.

The group exerting peer pressure can be in the workplace. As an employer you want the individual to become part of *your* group. It is worth a try — set up an atmosphere that will be conducive to helping the individual. Make them interested, listen to and accept their good ideas. All of these can be good management tools. It is to your advantage to have an employee who is sincerely interested in his or her work; this will tend to reduce dependence on chemical substances used as an escape from boredom or a disliked job.

This can also be true in financial terms to the company. If a person is nonproductive, it does not benefit the company. Increased absenteeism does not benefit the company. Insurance costs when there is an injury can be very expensive. Loss of production when the person is absent or on chemical substances can mean the difference between profit and loss.

Many individuals feel that peer pressure is always a negative; they feel that the individuals are being forced or coerced into doing something that they really do not want to do. It is felt that the individual would not have done something unless under the influence of peer pressure. Joining a group that uses drugs may lead the individual to drop out of religious activities or school, get drunk or high, experiment with increasing powerful types of chemical substances, or eventually perform criminal acts. But peer pressure is not always negative. Peer pressure, of itself, is neither positive nor negative; it is simply the way the group tends to influence the thinking and actions of individuals. When people join a group, there may be positive consequences. A child who joins a scout troop or youth group may discover sports, hobbies, or interests that will help him or her develop. Peer pressure from fellow members of a church or synagogue class might lead to deeper faith and increased social responsibilities.

We are mainly concerned with the person on our payroll who can be influenced and redirected by peer groups. One nebulous group that most people look at are the "adolescents" or young adults. Adolescence is a term to define the period of time between childhood and adulthood. But does society measure adulthood as an arbitrary age limit (18 or 21 years of age) or when a person demonstrates the qualities of adulthood: maturity, dependability, etc.? It is a hard question to answer since some individuals in the workplace may be chronologically adults but so immature that they can not function safely or productively. Adolescence may last into the middle or late twenties or even longer. During the transition from adolescence to adulthood, each person moves from having needs cared for by parents or others to a state in which the individual must supply his or her own needs; in other words, from a state of dependence to a state of self-reliance. This period is characterized by a struggle between the need for the security of parents to the need for personal independence. As an employer, you expect your employees to behave in a mature, logical, and "adult" manner. If they look like adults then they are presumed to be adults and act likewise. Unfortunately, some young people have only their childhood's limited experiences on which to base their judgments and behavior. Everybody experiences errors, frustrations, and painful experiences as adolescents. During this time of development, there are many haphazard experiences; some are filled with frustration or opportunity. It is a time when we seek, fail, have a sense of feeling inadequate, are lonely, and, importantly, possess a feeling of fear. These experiences help to develop reasoning powers and produce feelings that give us the basis for maturity and "adult" ways of perceiving. Many young people cannot handle these types of normal feelings and seek acceptance, love, etc., from their peers or from a substance that will replace and modify the feelings — drugs! When young persons are growing up, their world is their family. In adolescence, the individual has to begin to face the adult world. It is not easy — we would be fools if we thought it easy. During this developmental stage, the individual will question everything: family values, church values, and the laws of the land. They want to separate from family life. As employers or parents we tend to perceive this as rebellion against or rejection of societal values. This rebellion can be instigated or supported by the peer group. We see young people coming into the work force who want to demonstrate that they are no longer the baby of the family, but are yet unwilling to accept all the responsibilities that go with being an adult.

## ■ 7.3. MALADAPTED RESPONSE TO STRESS

Some people can respond to stress better than others. Most people accept that some stress is an unfortunate part of today's world (traffic, noise, rushed schedules, heavy work loads, ever increasing demands for productivity, etc.), but some others cannot handle all the stressors present.

When under real or imagined stress at work, at home, in financial matters, or from environmental conditions, most people can identify the stressors and adapt to the situation; but there are some people who can not adapt and will look for means to escape from the reality of the issues around them. Unfortunately, the misuse of chemical substances can often increase the level of stress due to fear of the police, fear of employers finding out, fear that there isn't enough money for the next drug purchase, stress from spouses or other loved ones suffering from the abuser's habit, or even a deep-seated realization that chemical substance use is wrong.

Misuse of drugs to relieve stress is clearly an inappropriate response; however, to the abuser, it seems logical to try to escape from stresses by taking chemical substances. Often this leads to a see-saw emotional state: taking drugs to relieve stress, then realizing that the chemical substance use is harming loved ones, and so forth.

## ■ 7.4. IS THE ABUSER "DIFFERENT"?

In spite of what some people may think, the typical substance abuser is often an average or even above average worker, at least at the beginning; has advancement potential; usually is friendly and well-liked; has technical competence; and may have a long record of company service.

Chemical substance abuse is usually a chronic and slowly developing illness. Some of the early indicators, which will be discussed further in the next section, are absenteeism, unusual excuses, mood changes, lower quality of work, long or irregular breaks, suspiciousness of others, excessive nervousness, and denial of having a problem. Later, the abuser may exhibit indicators such as spasmodic work pace, use of breath purifiers and cologne (especially with alcohol), financial problems, depressed condition, avoidance of supervisors and even fellow workers, increased occurrence of real illnesses, family problems, accidents on the job, and legal problems.

In short, at least at first, substance abusers are not really much different from any other employees. Until their abuse has begun to cause problems, they are very difficult to identify. Remember, however, that other illnesses or other problems can give symptoms exactly like those of substance abuse.

## ■ 7.5. IDENTIFICATION AND RECOGNITION OF THE SUBSTANCE ABUSER

One of the most difficult things an employer must do is to identify a substance abuser. A listing can be made of the various signs and symptoms of a chemically dependent person or substance abuser, but there are many medical conditions that can mimic the same signs and symptoms. Most employers are not medically trained and can misinterpret the signs and symptoms. One way of identification that is accepted by both management and labor organizations is to measure the worker's performance. Below is a checklist to help in identifying deteriorating job performance which may indicate an employee who is a substance abuser:

**Work schedule**

Using excessive sick leave, especially for minor conditions
Repeated unscheduled absences or tardiness, especially on Mondays or
 Fridays, before and after holidays or paydays
Leaving work early for a variety of reasons
Arriving late for work regularly

Leaving work area more than necessary with a variety of excuses
Taking excessively long lunch breaks

## Job performance

Alternating periods of high and low work productivity
Making more mistakes than usual
Making poor judgments/decisions
Missing deadlines
Wasting materials used on the job or damaging equipment
Slow at starting and completing tasks assigned

## Personality changes

Periods of alternating morale
Overreacting to criticism
Avoiding talking with supervisors regarding work issues
Difficulty in remembering directions or details
Difficulty in dealing with complex tasks or procedures
Requiring more time and effort than usual
Making inappropriate statements
Outbursts of anger, crying, or laughter at inappropriate times
Complaints from co-workers
More intolerant, resentful of fellow employees
Complaints from outside sources — the public, clients, other companies
Withdrawal from and avoiding fellow employees

## Indications of financial trouble

Regularly tries to borrow money from fellow employees
Wage garnishments

## Physical appearance

Deteriorating personal appearance/personal hygiene
Increasing nervousness and shakiness of hands or body
Changes in appearance after lunch or breaks

## Accidents

Increased incidents of accidents both on and off the job that interfere with
  job performance

## General

Any action or behavior that reflects discredit upon the individual, company
  or organization

Counseling is important in identifying and helping the substance abuser. Counseling may be defined as a systematic process of listening and communicating advice, instruction, or judgment with the intent of influencing a person's attitude and behavior. Counseling of employees is often difficult, especially when chemical substance abuse is suspected.

In fact, counseling is often not performed for many reasons, including lack of available time, lack of counseling skills, anxiety regarding discussing behavior, anxiety over one-to-one contact with potentially difficult people, and, last, fear of conflict. The purposes of counseling are many, including feedback to employee (positive as well as negative), identifying problems early, helping employees to become more self-reliant, evaluation of potential, and, ultimately, improvement of productivity. In order for the employer or supervisor to confront the employee on the existing or potential problem that is suspected, the following set of "do's" and "don'ts" has been developed.

### Do ...

Establish the level of work performance you expect from the employee
Set the minimum limits that you will accept
Determine what actions are acceptable and unacceptable to you
Document all absenteeism, tardiness, incidents on the job, and poor work
  performance; be specific with dates, times, and the people involved
Be consistent; treat all employees equally
Keep the discussion focused on facts
Base the confrontation on *work performance*, not on *personal issues*
Be firm; be direct; speak with authority; do not talk down to the employee
Be prepared to deal with the employee's resistance and denial, as well as
  hostility (discussing this with your Coordinator or Supervisor may help you
  deal with your own feelings and avoid a possible argument with the
  employee)
Discuss the company's position on treatment (Employee Assistance Program,
  if it is applicable)
Get a commitment from the employee as to what steps he or she will take
  to improve work performance
Make your expectations clear and specific
Continue to document
Emphasize that the company will exercise confidentiality
Explain that going for help does not exclude the employee from standard
  disciplinary procedures nor does it include special privileges

### Don't ...

Talk to the employee about personal problems or attempt to get involved in
  his or her personal life
Moralize or make value judgments
Ask "Why?" because this allows for reinforcing excuses or alibis
Allow the employee to box you into a corner; appropriate behavior and job
  performance are always the responsibilities of the employee
Let the employee play you against anyone else, including higher
  management; you are not in the middle here; the employee is!
Make idle disciplinary threats; if you give a warning, follow through with it
Be swayed or misled by emotional pleas, sympathy tactics, or "hard luck"
  stories; substance abusers are very good at manipulating people

Help an employee cover up his problem; you will only hurt them in the end
and delay the professional help he or she deserves

Be the victim of an employee's attempt to buy time, thus delaying the
inevitable

Reward the employee if they keep their problem hidden.

## ■ 7.6. DRUG TESTING

One of the most controversial issues confronting an employer is chemical substance testing. In recent cases, the courts have supported the
constitutionality of drug testing to order to identify substance abusers.
Chemical substance testing can provide the employer with a tool to use
to assist in maintaining a drug-free environment. It also allows the
employer the opportunity to identify the abuser who is in need of treatment, and, in some cases, monitor criminal offenders.

Drug testing may have an impact as a deterrent in that it is not
unusual for people seeking employment to refrain from drug use for
months until they get a job, and many workers fear the risk of having
their number come up in routine testing so they do not even experiment with drugs. The Americans with Disabilities Act (ADA) is assumed
not to apply; that is, the drug abuser is not considered protected as one
suffering from a disability.

Chemical substance testing has to be agreed upon by any labor
organization that represents the facility. Their support is needed and
vital to make this type of program successful. Joint management/labor
support will reduce any problems that may develop in having the employees participate. This will probably require a written management/
labor agreement.

Some companies have a chemical substance screening prior to
employment; others have periodic screenings. It is important that the
program be clearly defined to the employees. Completely random testing must be conducted and documented as to method used to identify
the employees for testing. Chain-of-evidence or custody is an important issue — everyone who handles the sealed specimen must sign and
protect the specimen; otherwise, in any legal action, you may lose the
case. Strict controls are necessary as to who obtains the sample, how it
was obtained, whether the specimen was temperature controlled, how
it was transported, and to whom in the laboratory the sample was given.
Who in the laboratory performed the test, and what method and equipment they used to conduct the test must be documented. The Federal
Government has a program for Federal employees prepared by the
National Institute on Drug Abuse (NIDA), Mandatory Guidelines for
Federal Workplace Drug Testing Programs.

## ■■ 7.7. SUMMARY

With the passage of the Drug-Free Workplace Act (text found in the Appendix of this book), employers have a mandated responsibility to identify and help those in their employ who abuse chemical substances. Chemical substance usage may result from peer pressure or response to stress or other reasons. Unfortunately, a circle is usually developed in which the abuser seeks release from stress or to be accepted, which leads to increased stress (fear of being arrested, fear of being found out by the employer, fear of being excluded from the peer group, etc.) which leads to increased chemical substance usage.

Identification of the abuser is not easy but an abuser may be recognized by a combination of actions or as result of chemical substance testing. The employer must always begin the process by understanding that the main purpose is to help the employee (even if he or she does not want the help) and only in extreme situations to fire or punish the abuser.

# 8 ■ EMPLOYEE ASSISTANCE PROGRAMS

## ■ 8.1. INTRODUCTION

The National Council of Alcoholism and Drug Dependence, Inc., (NCADD) has found the following:

- Alcoholism costs employers an estimated $33 billion in reduced productivity in 1988; other drug use cost an additional $7.2 billion in reduced productivity.
- Absenteeism among alcoholic or problem drinkers is 3.8 to 8.3 times greater than normal and up to 16 times greater among all employees with alcohol and other drug-related problems.
- Nondrinking members of alcoholics' families use ten times more sick leave as members of families in which alcoholism is not present.
- Of Chief Executive Officers responding to one survey, 43% estimated that the use of alcohol and other drugs cost them 1 to 10% of payroll.

Once an individual has been identified as a substance abuser or being chemically dependent, the company has but two options: to either assist or to dismiss the individual. It would be very easy to terminate the individual and send the problem on to another company, but, the better alternative is to help that individual. Retention of a trained employee is generally in the company's best interest since it is expensive to recruit and train personnel. How could saving an employee be accomplished? The company could contract with a local mental health unit and send its employees there, or, it could establish an in-house "employee assistance program".

What constitutes such an **Employee Assistance Program** (EAP)? Ideally, the program should be for all problems faced by the employees and not only for substance abuse or chemical dependency. There are many problems that affect why a person becomes a substance abuser: stress, financial difficulties, marital problems, children who may be substance abusers, etc. All of these can have an impact on a worker's performance. The EAP should be geared to handle all these conditions along with the dependency, which may itself be a symptom of other problems. An important aspect of the program is confidentiality. Confidentiality is a critical factor in making an EAP work. If the workers do not trust the counselors, all of your efforts will be lost. Many individuals are ashamed or reluctant to attend a program; therefore, it is necessary to keep their identities known only to the professional staff. If there is a leak about who is attending the program, the total program can be jeopardized. The program should be free with no hidden cost to the participants. Most insurance programs will pay for professional counseling of this nature. The Veterans Administration may also assist in payment to some veterans. Accessibility to the program must be on a 24-hour basis; there must be a professional on call to handle emergency situations. If you do not have a person available on your staff, arrange with a mental health unit or hospital in the area. Many critical situations will arise after everyone has left at the end of the normal working day or on holidays and weekends. The program must be for everyone in the organization; there should be no preferences for the various employees in the organization. The program should be all-encompassing for the employee's dependents since there may be other problems that need to be addressed and dealt with. Professionals need to run the program; this is not a program that can be run from the personnel or public relations departments. It requires the need of specialized health professionals to assist the employee. However, the personnel and public relations departments do have a role in training and making the program known to the employees. Particulars of the program must come from the professionals in the program to the individual shop stewards, supervisors, or members of union organizations as to how to get an individual referred into the program.

## ■ 8.2. WHAT IS AN EAP?

Ideally, any EAP program should be all-encompassing and based on the rationale that use of drugs, including alcohol, by employees in the workplace is totally unacceptable since it can adversely affect health, productivity, safety, security, public confidence, and above all — trust.

The EAP should not be limited to substance abuse or chemical dependency, but designed to cover a broad spectrum of personal problems encountered by employees. There are many problems that can cause a person to become a substance abuser: stress, financial difficulties, marital problems, children who may be substance abusers, etc. All of these can have a great impact on your workers' performance. The EAP should be designed to handle these problems along with the dependency problem.

The development of an EAP program for your specific situation should involve input from all aspects of your organization such as: labor relations, union, medical, security, and employee program assistance staff, whether they are within your organization or from a local mental health facility. Remember that an EAP is a formal, structured, systematic approach to assisting a troubled employee with the problems that are affecting job performance.

The EAP has to meet the following basic objectives:

- To motivate the employee to seek help before personal problems reach a severe or chronic stage that reduces the individual's ability to perform their job,
- To retain a valued employee,
- To restore the employee's productivity and enable him or her to lead meaningful and happy lives, and
- To provide the employee such help in a professional and confidential manner.

The program should be free with no hidden financial cost to the participants. Most insurance programs will pay for professional counseling of this nature. The military CHAMPUS program will also assist in payment for those eligible for benefits.

## ■ 8.3. TYPES OF EAPS

There are two ways for the company to handle an EAP program: internally or externally. The internal program gives the company the opportunity to define and manage the quality control of the program. With the external program, the company purchases services from a health care provider. Let us look at each program.

The internal program gives the company not only a means of quality control of the program but it can insure that the company goals and objectives are understood by the employee better than if the employee has to utilize a local health care provider. The relationship of the company staff to the employee is more emphatic than to someone who has no knowledge of the company. The group is part of the organization of the company and has good insight into the inner workings of the company. This advantage is important to the employee who works there.

The external program is one that has been established by a mental health facility/organization to assist companies who do not have the resources to develop an internal EAP program. Sometimes when a company cannot develop its own program due to the size of the company or the financial assets that are required to run the program, it will contract with a local mental health facility. While an external program utilizes professional resources which might be better trained and equipped, one problem with an external program is that it usually requires the employee to travel to the mental health clinic, something many people are reluctant to do. If possible, investigate the possibility that the external program personnel can locate an office physically in your facility.

## ■ 8.4. ELEMENTS OF THE PROGRAM

The EAP should consist, as a minimum, of the following elements: company policy; referral procedures; and voluntary participation procedures.

**Policy and procedures** — The company needs to adopt a written policy statement on alcoholism and substance abuse and any other problems covered by the EAP. This statement will be signed by management and labor representatives, where appropriate, and will reflect management's and labor's agreement to the EAP's objectives. It should be developed in consultation with the company's legal office or, in the case of a small business, the company's attorney. The policy statement should be distributed to all employees with information on how to clear up any misunderstandings and with whom to consult for information. It is important that all employees understand the policy.

**Referral procedures** — Written procedures for referral by either the employee or management to the EAP are critical. The procedures should explain how the program works. This will provide for assessment by the EAP committee or counselor; evaluation by professionals; referral for treatment; feedback from the referral center to management/labor; and any follow-up necessary to assist the employee.

**Voluntary participation** — Procedures need to be established for the employees to refer themselves to the EAP for assessment; evaluation by professionals; referrals for treatment; and follow-up. The EAP will notify management of employees who refer themselves.

One of the most important aspects of the program is its *confidentiality*. Confidentiality is a major factor in making any EAP work. If the workers do not trust the counselors, all of your efforts are lost. Without confidentiality, there will be no trust and your program will not work. Many individuals are ashamed or reluctant to attend a program; therefore, it is necessary to keep their identities known only to the professional staff. If there is a leak concerning who is attending the program, it can jeopardize the total program. The privacy of the individual must be maintained at all costs.

## ■ 8.5. ADMINISTRATIVE FUNCTIONS

In order for the EAP to function smoothly, there must be set into place certain administrative functions: where the EAP is on the organization chart; the physical location of the EAP; the record-keeping system; how the EAP fits into employee benefit plans; how contacts are made; education and training aspects of the program; involvement of management and labor in the program; resources; and evaluation of the program.

**Organizational position of the EAP** — The responsibility of the EAP should be positioned at an organizational level high enough to insure the involvement of top management and labor leadership in assisting and sustaining the program. In many cases, this means being attached to the office of the president or a vice president.

**Physical location of the EAP** — The EAP should be located so that it allows for easy and confidential access by the employees. This is often one consideration for an internal program.

**Record keeping system** — The EAP will have a record-keeping system carefully designed to protect the identity of the employee while facilitating case management and follow-up and providing a means to access statistical information to determine the efficacy of the program.

**Relation of the EAP to medical and disability benefit plans** — There should be a review of the company's medical and disability plan to insure or determine that the plans adequately cover appropriate diagnosis and treatment for alcohol, substance abuse, and mental health problems. Wherever possible, coverage should include both inpatient and outpatient care. Management and the employee should be familiar with provisions of the medical and disability benefit plans so they can advise clients clearly as to the extent, nature, and cost of the recommended treatment and reimbursement available.

**Counselor or committee point of contact** — There needs to be clearly identified a point of contact specifically trained to refer employees to professional services within the community. This contact can either be a counselor employed by the company or a committee made up of labor/management representatives.

**Education and training** — One of the critical elements of any EAP is the education and training of workers as well as their families in how the program works and what services are available. Personnel should be constantly updated on changes as they occur in the program by various educational techniques on its existence and availability. Information about the EAP should be made available to all new employees and their families during their orientation to the company.

There should be a major commitment by management to an ongoing educational program to inform workers on the hazards of the use and abuse of alcohol and chemical substances. Additional efforts should be made to educate employees about how to recognize an abuser and other problem areas.

**Involvement of management and labor representatives** — Management and labor representatives should be thoroughly informed of their key roles in utilizing the EAP services. Orientation for management and labor representatives should be updated on a regular basis.

**Resources** — The EAP should maintain current information about alcoholism and substance abuse treatment services and other resources available to the employee. Companies that have an external program under contract with a mental health facility will have current information on local resources, but those having an internal program should have information on: community health facilities (behavioral health centers, primary care clinics, state and private hospitals), Alcoholic Anonymous (Al-Anon), other self-help groups, and other health professionals.

The company must have regular contact with care providers to insure the current availability of the services.

**Evaluation** — There should be a periodic review of the EAP by both management and labor to ensure that each employee is receiving the most appropriate treatment available. This review should also provide an objective evaluation of the operation and performance of the program. Do not be afraid to make changes when needed.

## ■ 8.6. HOW THE PROGRAM WORKS

The EAP consists of several parts: identification and referral; employee acceptance into the program; follow-up; and accessibility.

**Identification and referral** — The EAP program should be available to the employees but is not a mandated program. The employee can take advantage of it. No one should be forced to use the program, and the program should offer the employee a variety of choices. Some of the choices are:

> Self-identification and referral to the EAP as a result of the general information campaign put forth by the company
>
> Identification and referral as a result of unsatisfactory work performance, since exposure to the general information campaign may not be enough to break through some employees' strong defenses against acceptance of their problem
>
> For these individuals, referral by their supervisor or union representative may be necessary.

**Acceptance of referral** — When an employee asks for a referral to the EAP, whether after confrontation or at an earlier stage where the employee realizes that he or she has a problem, the counselor or member of the joint labor/management committee will recommend the employee to the program for professional evaluation. The professional evaluator will diagnose whether or not the employee has a drinking or substance abuse problem and recommend the appropriate treatment program for the individual. At no time should the counselor or any member of labor or management attempt to perform these diagnostic functions.

**Follow-up** — The counselor will be responsible for follow-up work on the employee who is referred to the EAP for evaluation and treatment.

The counselor will meet with the employee at regular intervals to insure that the employee is in treatment and to assist the employee with any other problems that may be affecting treatment or the employee's social situation. For example, the counselor may have to make arrangements with the company to have the employee receive time off for treatment. This policy could be a part of the basic EAP program. The counselor will, at periodic intervals and without providing details, keep management or supervisor advised that the employee is still in treatment or implemented for a period of time following admission to the program.

**Accessibility to the program on a 24-hour basis and equal treatment** — There must be a professional on call 24 hours a day to handle emergency situations. If you do not have a person available on your staff, arrange with a mental health unit or hospital in the area. Many critical situations arise when everyone has left at the end of the normal working day or on holidays and weekends.

The program is a must for everyone in the organization, from the personnel in the board room down to the tool room. There should be no preferential treatment for any employee in the organization.

The program should be all-encompassing for the employee and his/her dependents. As we previously mentioned, there may be other problems that need to be addressed and dealt with; your EAP must be designed to accommodate all the employee's problems. You may solve one problem, but if you do not look at the other problems the individual has, you may not have accomplished your goal of rehabilitating the individual back into the workforce.

Professionals need to run the program, as this is not a program that can be run from the personnel or public relations departments. It requires specialized health professionals to assist the employee. The personnel and public relations departments do, however, have a role in the program to train the individual shop stewards, supervisors, or members of union organizations as to the means for referring an individual into the EAP program.

## ■ 8.7. FIVE-STEP PROGRAM TO ASSIST THE EMPLOYEE

Helping your employees, or yourself, involves a stepwise process consisting of recognition that there is a problem, documentation for legal purposes, action to help, referral to an EAP, and reintegration of the employee back into productive employment.

### Step Number 1: RECOGNITION

This is discussed in Chapter 7 on recognizing that a problem may exist and not turning your back on the problem, as the problem will normally get worse instead of better. Recognizable on-the-job signs are:

**Employee**
Tardiness
Physical appearance
Absenteeism
Relationship with fellow employees
Attitude and mood
Poor work habits
Recurring illnesses
Accidents or alleged illnesses
Low productivity
Relationships with significant others

**Supervisor**
Becomes lax in his supervisory duties
Issues conflicting instructions to employees
Uses employees' time and skills to cover responsibilities clearly within his
 job description
Submits incomplete reports and data
Mismanages budgets
Fails to coordinate schedules

## Step Number 2: DOCUMENTATION

Keep an accurate and up-to-date file of the employee's work performance. Without it, the supervisor doesn't have a leg to stand on and can turn an otherwise effective confrontation into a case of a "your words against mine" type situation. Providing "proof" helps the employee to become more aware and to comprehend the scope of the problem. Be specific and objective! It is very important that the data that are collected be as specific as possible and must be centered on job performance or any unusual behavior on the job. Recurring patterns ought to be noted. Do not go on hearsay. Everyone has an "off day" once in a while, so observations or documentation should go over a period of time. Collection of data helps the supervisor make a fair and impartial assessment of the employee's job performance. It also guards against "euphoric recall", that is remembering the peaks of performance — the "good days" and not the valleys or "bad days". The supervisor is not a counselor or judge of the employee. Rather, he or she is someone who assesses performance and then asks for assistance. Follow a formal disciplinary process consisting of informal verbal conferences, written conferences, written warnings, probation, suspensions, termination. Establish a time table and follow through with appropriate ACTION.

## Step Number 3: ACTION

Follow a formal disciplinary process consisting of, in order: informal verbal conferences, written conferences, written warnings, probation, suspensions, and termination. The purpose of these actions is to notify the employee that you recognize there is a problem and that you want to help. One of the first steps in your action is to refer the employee for professional assistance.

**Step Number 4: REFERRAL**
Refer the employee to your EAP coordinator, or EAP program if you do not have an in-house program, preferably as soon as the problem has been identified, for the longer you wait, the employee may not benefit from treatment. Don't refer the employee as the last straw before termination. Make a positive referral early enough for intervention to help before it's too late.

**Step Number 5: REINTEGRATION**
This step involves the supervisor helping the employee to become re-adjusted to the work environment following treatment. For example: an employee who has been in an alcoholic or drug rehabilitation program. This is a very important step in that it will demonstrate to the employee that the company and people do care for his/her well-being.

## ■ 8.8. COST EFFECTIVENESS OF AN EAP

The National Council of Alcoholism and Drug Dependence has surveyed industry to determine the effectiveness of EAPs. It should be remembered that rehabilitating a valuable employee cannot be measured only in dollars; however, the NCADD results are as follows:

- For every dollar they invest in an EAP, employers generally save anywhere from $5 to 16. The average annual cost for an EAP program ranges from $12 to 20 per employee.
- General Motors Corporation's EAP saves the company $37 million per year, or $3,700.00 for each of the 10,000 employees enrolled in the program.
- United Airlines estimates that it has a $16.95 return for every dollar invested in employee assistance.
- Northrop Corporation reported a 43% increase in productivity of each of its first 100 employees to enter an alcohol treatment program. After three years' sobriety, the average savings for each employee was nearly $20,000.
- Philadelphia Police Department employees undergoing treatment reduced their sick days by an average of 38% and their injured days by 62%.
- Oldsmobile's Lansing Michigan plant saw the following results in the year after its alcoholic employees underwent treatment: lost man hours declined by 49%, health care benefits by 29%, leaves by 56%, grievances by 78%, disciplinary problems by 63%, and accidents by 82%.

## ■ 8.9. NATIONAL INSTITUTE ON DRUG ABUSE ASSISTANCE TO EAPS

The National Institute on Drug Abuse (NIDA) has a toll-free number (800-843-4971) to provide individualized technical assistance for businesses, industry, and unions to assist in developing and implementing a comprehensive drug-free workplace program. Corporate executive officers (CEOs), union representatives, and managers responsible for corporate policy are encouraged to call the toll-free number for assistance in assessing their programmatic needs and to help to prepare their organizations to deal with current or potential problems caused by drugs in their workplace. This goes hand in hand with developing and maintaining an EAP.

The NIDA program will do the following:

- Assess the nature and extent of drug abuse in the organization
- Develop and implement a drug abuse policy
- Choose an employee assistance program (EAP) model that is compatible with the individual organizations, including organizational benefit programs
- Implement employee education and supervisory training
- Evaluate the effectiveness of a drug abuse program in terms of cost and human factors
- Understand the technical, legal, and employee relations aspects of drug testing
- Identify signs and symptoms of drug abuse

While roughly 90% of the Fortune 500 companies have established EAPs, this percentage is much lower among smaller companies. Only 9% of businesses with fewer than 50 employees have EAP programs. Of all U.S. businesses, 90% fall into this category.

## ■ 8.10. SUMMARY

The National Institute on Drug Abuse (NIDA) and the Alcohol, Drug Abuse, and Mental Health Administration (ADAMHA) estimate that chemical dependency costs nearly $100 billion dollars in lost productivity each year. Each alcoholic/drug-dependent employee costs his/her employer an amount equal to at least one quarter of the annual salary in hidden losses such as absenteeism, tardiness, spoiled material, excess scrap, unsatisfactory production, medical bills, and accidents. The problems of substance abuse are private problems until they affect safety, productivity, and the cost of running a business. The symptoms may show up as work performance declines over a period of time; the losses associated with decreased production, excessive absenteeism, accidents, conflicts with work mates, and disciplinary action add up.

As an employer, once you have an individual who has been identified as a substance abuser or as being chemically dependent, you have but two options: to assist or to dismiss the individual. A company can assist its employees by establishing an "employee assistance program" (EAP).

The EAP can be either internal and external. The internal program allows the company to manage the entire program. With the external program the company purchases services from a health care provider.

The program to assist your employees usually has five steps: recognition, documentation, action, referral, and reintegration.

To assist employees with substance abuse or other related problems, a company can develop an employee assistance program. This program must be all-encompassing to ensure that complete services are available to all employees and their families. A valued employee deserves the chance to again become a productive employee!

# 9 LAWS AND REGULATIONS GOVERNING SUBSTANCE ABUSE

## 9.1. INTRODUCTION

Recognizing that chemical substance abuse is a harmful activity, there have been many laws and implementing regulations passed by the federal, state, and local governments in an effort to reduce (or hopefully, to eliminate) the problem from society.

## 9.2. EARLY LEGISLATION

The National Drug Import Law of 1848 was enacted to control the quantity and purity of drugs coming into the U.S. The law gave the government the right to reject drugs that did not meet medical standards of quality and purity.

The Opium Act of 1887 was passed by Congress to execute an agreement with China that forbade the importation of opium into the U.S. by Chinese nationals and forbade Americans from engaging in the Chinese opium trade.

However, even with these laws, one of the major problems confronting our nation during the early 1900s was addiction to opium and opiates. The federal government passed the first Pure Food and Drug Act in 1906; this particular piece of legislation was designed to deal with opium addicts and to control the patent medicine industry. The 1906 act required that patent medicines indicate on their labels or bottles if the medicine contained an opium-based ingredient. Later legislation went on to require the manufacturer to list the actual quantity of each drug in the mixture. This is very similar to the legislation of today which requires manufacturers of most products to list all the ingredients and quantities. The act also specified that drugs had to meet certain purity and quality standards.

The Opium Exclusion Act of 1909 prohibited the importation of opium or its derivatives except for medicinal purposes. The law was regulated by the Secretary of the Treasury. Some social scientists believe that this law also protected the addict by providing quality control for opium products.

It was not until 1914 that the Harrison Narcotic Act was introduced; this Act sought to cut off the supply of previously legal opium-containing products. Obviously, this had a great impact on those who were addicted. The addicts had to find other sources of opium, now illegal, and the quality and purity of these other drugs did not meet the quality control that the government had earlier imposed. These illegally obtained drugs were often contaminated with other materials, changed by adding other chemicals, or even replaced with anything that could be found, just to sell to the addicts.

The U.S. Senate considered instituting a new regulation that would reduce, but not eliminate, the amount of opiates imported into this country. The major proponent for a total ban, however, was Secretary of State William Jennings Bryan, a man who had very strong feelings on this subject and who personally urged that a law banning opiates be instituted.

The Harrison bill was not really an act to reduce the number of those addicted, but a tax act as titled: "An Act to provide for the registration of, with collectors of internal revenue, and to impose a special tax upon all persons who produce, import, manufacture, compound, deal in, dispense, sell, distribute, or give away opium or coca leaves, their salts, derivatives, or preparations, and for any other purposes." This law, in effect, called for the taxation of pharmaceutical companies, importers, pharmacists, and any physician that produced or dispensed the drugs. The irony of this bill is that patent medicine manufacturers were exempt! They could produce and sell a patent medicine if it contained "less than two grains of opium, or not more than one-fourth grain of morphine, or not more than one-eighth grain of heroin... in one avoirdupois ounce."

These laws did not stop the use of opiates, but did put a tax and certain restrictions on them. Physicians, for instance, were only required to keep records of the various drugs and medicines that they prescribed or issued. This law was very nebulous, and many physicians continued to prescribe opiates for their addicted patients. This presented a legal problem for law enforcement personnel and the judiciary. Was addiction a disease? Law enforcement and judicial personnel felt that it was not, while many physicians felt that it was. Physicians could "in the course of his professional practice" dispense any medication that they felt was beneficial to their patients. Many physicians were arrested for violating the law, and some were convicted. Even if the physician were exonerated of the charges, the notoriety and trial would generally ruin his or her career.

The Volstead Act of 1920 curtailed the production and sale of alcohol. All nonmedical uses of alcoholic beverages were prohibited. This brought about the period known as Prohibition, which ended when the Volsted Act was replaced by legalization with strict controls and taxation of alcohol-containing beverages.

The Jones-Miller Act of 1922 established firm penalties for violation of the Harrison Narcotic Act of 1914. In 1922, Congress passed amendments to the Harrison Act which banned the importation of heroin into this country even for medical reasons. Many of the morphine addicts of this time did not stop their conversion over to heroin. Heroin remained a drug of choice even after the ban was instituted.

The Marijuana Tax Act of 1937 brought marijuana under stern controls similar to those regulating the use of opiates.

The Food, Drug and Cosmetic Act of 1938 was a beginning for control of the quality and purity of food, drugs, and cosmetics shipped via interstate commerce.

The Opium Poppy Control Act of 1942 was signed by President Franklin D. Roosevelt and required that growers of opium poppies be licensed by the Secretary of the Treasury.

The Boggs Act of 1951 set up a series of graduated mandatory sentencing procedures for narcotic drug offenses. Subsequent to the Boggs Act, many state legislatures enacted what were called "little Boggs acts", many of which established strict mandatory sentences. Some judges have refused to hear cases in which there is mandatory jail time since it limits their control over the case and does not allow for mitigating circumstances.

## ■ 9.3. RECENT LEGISLATION

The Narcotic Drug Control Act of 1956 was even more punitive than the Boggs Act of 1951; however, it did differentiate among drug possession, drug sale, and drug sales to minors. The medical use of heroin was prohibited under this act.

Unlike many of the earlier (and even later) laws, the Narcotic Addict Rehabilitation Act of 1966 viewed narcotic addiction as being symptomatic of a treatable disease and not a criminal condition.

The Drug Abuse Control Amendments of 1966 brought sedatives, stimulants, and tranquilizers under tighter controls. Hallucinogens were specifically added to the laws. Enforcement became the responsibility of the Bureau of Drug Abuse in the Food and Drug Administration (FDA).

The Comprehensive Drug Abuse Prevention and Control Act of 1970 increased enforcement activities and provided stiffer penalties for dealing in cocaine and a number of other dangerous drugs. It replaced previous legislation on the control of narcotics, marijuana, sedatives and stimulants and placed their control under the Department of Justice. Under this Act, drugs are classified into five schedules according to their potential for abuse and therapeutic usefulness; these schedules are discussed in a later section of this chapter. First-time illegitimate possession of any drug was considered a misdemeanor and penalties were reduced. Provisions were made for rehabilitation, education, and research. This act introduced the "no-knock" house search, making it legal for law enforcement agencies to enter a house with a probable cause when it is thought that drugs are being used.

Executive Order 12564 of September 5, 1986 mandated that federal agencies formulate a drug-free workplace plan which would include drug testing, among other components including education.

Public Law 100-71, Supplemental Appropriations Act, July 11, 1987 specified the conditions under which Executive Order 12564 would be implemented.

Mandatory Guidelines for Federal Workplace Drug Testing Programs, *Federal Register*, April 11, 1988, set forth the standards for all aspects of drug testing and procedures in the federal government program.

Department of Defense Regulations, *Federal Register*, September 28, 1988, required Defense contractors to institute and maintain a drug-free workplace program.

Department of Transportation Regulations, *Federal Register*, November 14, 1988 required drug testing of the more than 4 million employees in safety and security jobs in all major modes of transportation.

Nuclear Regulatory Commission, *Federal Register*, June 7, 1989 required licensees authorized to operate or construct nuclear power reactors to implement fitness-for-duty programs to include drug testing and education.

## ■ 9.4. PHYSICIAN LICENSING AND PRESCRIPTION SYSTEM

Upon graduation from an accredited medical school, a physician is still not allowed to independently practice medicine until he or she passes the state medical board's examination. Upon passing and being granted a license to practice medicine in that state, the physician is judged to have met all requirements established to prevent unethical and immoral persons from calling themselves medical doctors. With the state license goes the right to prescribe any drugs which, in the physician's opinion, are necessary for the patient's well being. With a few exceptions for drugs which are considered as experimental, a physician is free to prescribe any medication. And, even in the case of many experimental drugs, the physician can prescribe the drug as long as he or she is aware that there could be legal complications should anything go wrong, since the physician could be considered as not following accepted practice. All states have laws which place requirements of graduation and licensing on physicians, nurses, and pharmacists. Any violation of these laws can mean loss of license (and possibly a lawsuit or imprisonment), so most medical personnel adhere strictly to the codes of ethics of their profession and to the laws under which they practice.

In order for a controlled drug to be legally dispensed, the attending physician must write a prescription which must be taken to a licensed pharmacist, who then may dispense the drug, but only in the amounts and strengths specified on the prescription. In some cases of noncontrolled drugs, the pharmacist can make substitution of the generic drug in place of the brand name drug. The pharmacist is required to keep records of the quantity of each drug received and dispensed along with all prescriptions. Refills of controlled drugs must be justified. The U.S. prescription system is one of the most carefully controlled in the world, as anyone who has traveled to certain other nations can attest.

Ethical drugs, those produced by legal drug companies, are carefully evaluated both for efficacy and for potential harmful side effects. Details of the licensing of drugs are discussed later in this chapter.

## ■ 9.5. ROLE OF THE FOOD AND DRUG ADMINISTRATION (FDA)

The Food and Drug Administration is charged with ensuring the safety and efficacy of all legal drugs produced in the U.S. Before any drug can be offered for sale, it must undergo extensive testing both by the manufacturer and by watchdogs within the federal government.

## ■ 9.6. DRUG SCHEDULES

The Schedules of the Comprehensive Drug Abuse Prevention and Control Act of 1970 are

### Schedule I Substances
Drugs in this schedule are those drugs that have no accepted medical use in the U.S. and have a high abuse potential. Some examples are heroin, marijuana, LSD, peyote, mescaline, psilocybin, the tetrahydrocannabinols, ketobemidone, levomoramide, racemoramide, benzylmorphine, dihydromorphine, morphine methysulfonate, nicocodeine, and nicomorphine.

## Schedule II Substances

The drugs in this schedule have a high abuse potential with severe psychic or physical dependence liability. Schedule II controlled substances consist of certain narcotic drugs and drugs containing amphetamines or methamphetamines as the single active ingredient or in combination with each other. Examples of Schedule II controlled drugs are opium, morphine, codeine, hydromorphone, methadone, pantopon, meperidine, cocaine, oxycodone, anileridine, oxymorphone; and straight amphetamines and methamphetamines. Also in Schedule II are phenmetrazine, methylphenidate, amobarbitol, pentobarbital, secobarbital, and methaqualone.

## Schedule III Substances

The drugs in this schedule have an abuse potential less than those in Schedules I and II and include compounds containing limited quantities of certain narcotic drugs and nonnarcotic drugs, such as: derivatives of barbituric acid, except those that are listed in another schedule, glutethimide, methyprylon, chlordexadol, phencyclidine, sulfondiethylmethane, sulfonmethane, nalorphine, benzphetamine, chlorphentermine, chlortermine, mazindol, and phendimetrazine. Paregoric is in the schedule as well.

## Schedule IV Substances

The drugs in this schedule have an abuse potential less than those listed in Schedule III and include such drugs as: barbital, phenobarbital, methylphenobarbital, chloral betaine, chloral hydrate, ethchlorvynol, ethinamante, meprobamate, paraldehyde, pentaerythritol chloral, methohexital, fenfluramine, diethylpropion, and phentermine.

## Schedule V Substances

The drugs in this schedule have an abuse potential less than those listed in Schedule IV and consist of preparations containing moderate, limited quantities of certain narcotic drugs, generally for antitussive and antidiarrheal purposes, which may be distributed without a prescription order.

■■ 9.7. ROLE OF THE DRUG ENFORCEMENT
      ADMINISTRATION (DEA)

The Drug Enforcement Administration has the responsibility of enforcing legislation as well as of curbing the flow of drugs into the U.S. The DEA is charged with reducing the flow of drugs entering the U.S. and investigating drug diversions. Drug diversions are drugs taken from a legal source and diverted into an illicit channel.

# ■ 9.8. DRUG-FREE WORKPLACE ACT OF 1988

One of the most important pieces of legislation that impacts on the workplace, other than the Occupational Safety and Health Act of 1970, has been the Drug-Free Workplace Act, passed by the 100th Congress in 1988.

This law prohibits the manufacture, distribution, possession, and use of controlled substances in the workplace. The act covers the following substances:

- Cocaine, heroin, morphine, and other narcotics
- Stimulants, (amphetamines, etc.)
- Depressants (barbiturates, tranquilizers, etc.)
- Hallucinogens (LSD, PCP, etc.)
- Marijuana

Alcohol is excluded from the Act, but most employers include it in their own drug-free workplace policies.

The Act affects those companies that receive any federal grants or federal contracts in excess of $25,000.00; it also affects most government employees. The law was enacted to protect everyone from the dangers of drugs used in the workplace. When an employee uses chemical substances, whether they be illicit or even many over-the-counter medications, there are concerns of safety, productivity, security, and public trust.

Safety is compromised because the individual does not have full control of his/her systems. This can lead to an increase in accidents, injuries, and even deaths. Productivity is jeopardized with the decline in quality of work, output, and an increase in the days absent because of the drug usage. Security is also impaired because the drug user may steal to keep up the habit. Users may steal equipment, money, or even company trade secrets to sell to have money to buy drugs. Public trust will be in jeopardy if drug usage is found to be the major factor in a serious or even fatal accident. For instance, when a train wreck occurred in the Northeast that killed several people, and it was subsequently found that the engineer of the train was under the influence of drugs, legal liability was greatly increased.

Not only do drugs affect the workplace by jeopardizing the items above, but they also do the following:

- Harm workers' health
- Interfere with the worker's judgment and coordination
- Reduce the span of attention
- Decrease worker performance
- Impair relationships with friends and family
- Increase arguments among workers, fellow staff, and management

Primarily, the Act is intended to:

- Educate the work force in the affects of the use of drugs
- Make the workplace a safe and healthy place
- Reduce the number of accidents by improving safety
- Cut costs to the employer, by reducing accident costs, theft, and insurance claims
- Improve understanding of why and how people use drugs

## ▆ 9.9. REQUIREMENTS UNDER THE ACT FOR EMPLOYERS

The Drug-Free Workplace Act requires that the employers do the following:

Have and publish for all employees to read and understand a policy statement that prohibits use, manufacture, sale, and possession of drugs in the workplace.

Establish a drug awareness program and, in that program, cover the following topics:

1. The dangers of using drugs, and what they do to your body
2. That there is in place an employers' policy. If you are caught with drugs what will the consequences be — dismissal?
3. That employees shall give notice that they understand the Act and are aware of the rules incorporated by this Act
4. Employers have to make a "good faith" effort to try to make their workplace completely free of drugs
5. That require personnel who have had a drug-related conviction are required to be enrolled in an employee assistance program (EAP) or other rehabilitation program, or even to face penalties
6. That an employer dealing with the federal government, must inform the contracting officer's representative (COR) within 10 working days if a violation occurs.

All of this is important to the employer if his employees fail to read and understand it, or if the employer does not enforce the Act, since he may lose a contract or grant, and, in some cases, may even be barred from future contracts or grants for up to a period of 5 years.

## ▆ 9.10. REQUIREMENTS UNDER THE ACT FOR EMPLOYEES

Employees have to take part in the program by having an understanding of the company's policies against drug use and its work policies, and knowing the criminal and civil penalties for the possession and sale of controlled substances. They must also report to the employer any conviction on a drug-related charge within five (5) days. This can range from a motor vehicle violation e.g., (DUI) to an actual charge of drug possession.

One of the purposes of this book is to help make each employee aware of the dangers of drug abuse and how drugs can affect the body, mind, and society. This book can be an important part of an employer's education program to comply with the Federal Act.

## 9.11. SUMMARY

Upon being granted a license to practice medicine, a physician is judged to have met all requirements to legally prescribe any drugs which are necessary for their patient's well being. States have laws which place requirements of licensing of physicians, nurses, and pharmacists.

For a controlled drug to be legally dispensed, the physician must write a prescription which is taken to a licensed pharmacist, who then may dispense the drug as specified on the prescription. The U.S. prescription drug system is efficient and well run.

Drugs produced by legal drug companies are carefully evaluated both for efficacy and for potential harmful side effects.

The Drug-Free Workplace Act of 1992 affects those companies that receive any federal grants or any federal contract in excess of $25,000.00; it prohibits the manufacture, distribution, possession, and use of controlled substances in the workplace and covers: cocaine, heroin, morphine, and other narcotics; stimulants (amphetamines, etc.); depressants (barbiturates, tranquilizers, etc.); hallucinogens (LSD, PCP, etc.); and marijuana.

The Drug-Free Workplace Act requires that the covered employers have and publish a policy statement that prohibits drugs in the workplace, and establish a drug awareness program.

Employees have to have an understanding of the company's policies against drug use. They must report to the employer any conviction on a drug-related charge within 5 days.

Although many pieces of legislation have been promulgated to protect the public of the U.S., there is still availability of illicit and licit substances being abused in the American workforce. Strict penalties and education are a vital factor in attempting to curb the influx of drugs into our country. There is no easy answer to the problem; it will take a concerted effort to eradicate the problem.

# 10 SUBSTANCE ABUSE IN SPECIFIC WORK SETTINGS

## 10.1. INTRODUCTION

Although chemical substance abuse cuts across all ethnic groups, socioeconomic classes, and occupations, certain work settings stand out because of importance to the community, easy access to chemical substances, cataclysmic results from accidents, or public perception. Some of the more visible work settings will be discussed in this chapter.

## 10.2. SUBSTANCE ABUSE IN THE MEDICAL AND ALLIED SCIENCES PROFESSIONS

Health care providers are relied upon to save lives and make sick people well; most do so daily without any recognition. However, a few in the medical community find that chemical substances are readily available to relieve perceived pressures of their work. Often co-workers feel that it is right to cover up for someone suffering from addiction; unfortunately, this leads to delay in treatment. Surprisingly, chemical dependence among health care workers is higher than for the general population. Many health care workers are able to cover up their dependency for a long time. But no one ever gets "better" on their own when chemical dependency is the problem. Some of the general signs of chemical dependency are discussed elsewhere in this book; however, specific signs in the health care worker include sloppy handwriting and unclear or rambling notes on a patient's chart together with ordering or delivering the wrong medicines. Serious injuries or even death can result from impaired health care workers making mistakes. Access to drugs can lead to dependency; some doctors and nurses have been discovered ordering unnecessary or excessive medications for patients in order to steal the drugs before they get to the patient.

The most commonly abused drugs among health care providers include alcohol, of course, but also prescription drugs, especially tranquilizers and other depressants in order to relax. The use of alcohol during work hours is not very common among health care workers since the smell of alcohol would indicate its use. Some personnel use stimulants to keep awake during long shifts (like the eleven-to-seven) and extended work hours (such personnel as interns and residents).

While it would seem that doctors, nurses, and other care givers would see the dangers of chemical substance misuse in emergency rooms and elsewhere, many such workers feel that they are "better" or "smarter" or "know more" so they cannot become addicted. Unfortunately, drugs do not recognize advanced degrees and training. The availability of drugs for doctors, nurses, and pharmacists can be quite tempting.

## ▬ 10.3. SUBSTANCE ABUSE IN THE TRANSPORTATION INDUSTRIES

Those who move the passengers and essential materials of today's world are vitally important to society; unfortunately, a perceived need for haste in transport has led to problems in misuse of chemical substances. We are all aware that shipping companies try to constantly outdo their competitors with promises of speed of delivery. Truck drivers are constantly reminded that speed is important. In addition, many truck drivers are paid based upon number of miles traveled; this leads to drivers overextending themselves to continue driving even when tired in order to make a few more miles. While the U.S. Department of Transportation (DOT) has regulations limiting the number of miles a commercial truck driver may drive, a few drivers have found ways around the record-keeping so as to exceed the maximum. Recently, requests have also been made to increase the number of miles allowed under the regulations; what effect, if any, this may have on chemical substance misuse is not known.

Chemical substance abuse among airline pilots appears to be quite rare due to the on-going requirements for fitness testing and maintenance of demonstrated skills. Chemical substance testing has been a topic of discussion between the airlines and the pilots' union, but does not seem to be a major problem for the commercial airlines as yet. The greater danger lies in the noncommercial or private pilot who has misused a medicine, even an over-the-counter medication, with resulting lack of attention to details and slowed responses. All pilots, commercial and private, need to be constantly aware of the effects of even seemingly harmless medications on perception, attention, and alertness.

Railroad workers unfortunately have been responsible for serious accidents and deaths while under the influence of chemical substances. Several accidents which resulted in deaths occurred when the engineers were under the influence of drugs. The most common drugs found following rail accidents are alcohol and marijuana, although cocaine has been found in a few cases. Testing of locomotive engineers is controversial but may become the norm in a few years.

Although it is more commonly known than many similar situations, the case of the *Exxon Valdez* is not unique. The use of alcohol appears to be fairly common among crews of ships. This is partially due to the long, lonely hours spent at sea away from families and loved ones, unlike most others in the transportation industries. The result of such alcohol abuse can be seen in the damage to the ship and the resulting environmental damage done by the *Exxon Valdez* oil spill in Prince William Sound, Alaska.

The most commonly abused chemical substances in the transportation industries are alcohol and stimulants. In their desire to keep moving, some truck drivers and railroad engineers rely on chemical substances to keep awake; almost any truck stop or rail yard has such chemical substances available in some degree. Those using these chemical substances don't look upon this as a problem; they are simply trying to remain alert for longer periods of time so as to be more efficient. The major trucking companies have programs to identify and help the chemical substance abuser; however, many drivers are independent business people who own their own rigs and therefore are not covered under chemical substance testing or other programs. The railroads likewise have active programs to identify and help those who misuse chemical substances.

Beer is quite commonly consumed by truck drivers, railroad engineers, and ship crews. In moderation, such usage may not be a problem, although some people incorrectly think of alcohol as a stimulant which may help keep them awake, while it is really a depressant which can cause drowsiness and sleepiness. Stimulants such as amphetamines are fairly common among long-haul truck drivers who abuse chemical substances. The combination of alcohol and stimulants is particularly dangerous since they have mutually opposite effects and one might increase the effect of the other.

The use of depressants is apparently not very common among truck drivers while on the job; there is some evidence that they might be used to "bring down" the driver upon arriving at a rest stop or ultimate destination. Such "see-saw" chemical abuse can cause physical and mental damage to the abuser.

Chemical substance abusers among transportation employees most commonly use chemical substances to keep awake during long hours of monotonous driving, long nights at sea, and to increase the number of miles traveled. Unfortunately, while still not a problem of every truck driver or engineer or sailor, chemical substance misuse is fairly common, especially among younger workers and those who feel that they have to constantly remain alert. Labor unions, fearful of possible retaliation against some drivers and excessive punishments by companies and possibly overzealous police, have not strongly supported mandatory chemical substance testing, and a few have opposed it, apparently based on sufficient cause requirements under the Constitution whereby a law enforcement person must have some reason for suspecting a person of wrongdoing before stopping him/her. These problems will have to be worked out in management-labor negotiations and in the courts of law. For now, the use of chemical substances by drivers is a potential danger to other drivers and themselves. Mandatory chemical substance testing would go a long way to removing the dangers of having an overtired or "spaced-out" truck driver on the road, possibly hauling hazardous chemicals or other dangerous cargoes.

The U.S. government has issued regulations under 46 CFR Parts 4 to 35 for the commercial merchant marine fleet. These regulations specify conditions under which mandatory drug testing will be performed, either following an accident or routinely in the case of ship's officers and crew. The requirements are forced by the statutory requirement that all seamen be mentally and physically able to perform their duties. Detailed descriptions may be found in 46 CFR Part 16. In general, they require that testing be conducted following any "marine casualty, discharge of oil..., or discharge of hazardous substance...." Refusal to take part in the drug testing can result in suspension or termination of employment. Random, routine testing is to be conducted and results made available to the U.S. Coast Guard. The drugs to be tested for are listed in 46 CFR 16.350 using the test methods in 46 CFR 40.24. All commercial maritime employers of more than 50 persons are required to implement the provisions of the regulations, to include an employee assistance program.

## ■ 10.4. SUBSTANCE ABUSE IN THE MINING AND PETROLEUM INDUSTRIES

Mining is a dangerous profession under the best of circumstances, as is work on drill rigs and in oil fields where there is potentially hazardous machinery and a less-than-ideal working environment. Miners have to work with complex mining machinery and work in an environment that may consist of noise, dust, fumes, potential for explosions or rock falls, and poor ventilation and lighting.

Having someone impaired by drugs in this setting jeopardizes their and their workmates' lives. The U.S. Department of Labor's Mine Safety and Health Administration (MSHA) first addressed the problem of the impaired miner in 1986, forming a Mining Industry Committee on Substance Abuse comprised of union, mine, state, and MSHA representatives. This committee, in conjunction with the National Mine Health and Safety Academy, Beckley, West Virginia, developed programs and resources on how to effectively deal with the impaired miner.

There are few readily available statistics as to the number of miners, office workers, or managers in the mining industry who use alcohol or drugs. Many consultants to the mining industry agree that there is a parallel to reports by the media that drug use in the American workplace is a serious problem and that the situation in the mining industry is no different.

The MSHA position is not to tolerate substance abuse in the mining industry. They highly encourage the use of EAPs and education to prevent miners from using drugs. MSHA has requirements (30 CFR) that contain provisions pertaining to coal, metal, and nonmetal mining under which persons engaged in potentially dangerous practices in a mine under the influence of alcohol or drugs can be promptly removed from the premises. Many mining companies have specific rules to prevent their employees from using alcohol or drugs at the mine site, but penalties for breaking the rules vary widely from one company to another. Findings that drug use goes from the "tool room to the board room" are very similar to other professions.

For the mine worker who has a serious alcohol or drug problem, the effectiveness of the company's employee assistance program may be the difference between keeping his job or losing it, and the program may have an impact on all of his or her personal relationships. Even more critically, effective counseling by company or outside specialists may help the employee to avert a tragedy some day in the form of a fatal or disabling mine accident or other traumatic incident, even off the job, in which substance abuse has a role.

## ■ 10.5. SUBSTANCE ABUSE IN THE CONSTRUCTION INDUSTRIES

The construction industries encompass a wide variety of jobs, ranging from minor alterations of existing structures to building large buildings and expansive paved areas. Therefore, the types of drug abuse appear to vary widely. Those working on small jobs, usually consisting of a crew of only a few workers, seem most prone to abuse of alcohol. The use of other drugs might be small among the more experienced workers; however, the apprentice workers and day hires seem to be more likely to abuse marijuana in addition to alcohol. The use of other drugs, such as cocaine and heroin, does not seem to be a major problem in most parts of the country. Even the use of tranquilizers and stimulants does not seem to be very common, especially among the smaller contractor employees. Larger jobs, and those in larger communities, appear to more closely correspond to drug abuse in their area; that is, there seems to be more use of hard drugs on large job sites, probably due to the fact that supervisors and foremen are less likely to know the workers as well, and neighborhoods near such sites might be more likely to have pushers. Interestingly, use of cocaine and heroin seems to be less associated with construction accidents; it is possible that neither of these are drugs of choice among most construction workers. Many construction workers seem to enjoy alcoholic beverages, especially beer and ale; most do not consume to excess, although the few who do can place co-workers at risk of accidents. Wines do not seem to be very popular among construction workers, possibly due to peer pressure.

There are many hazards associated with construction workers being on drugs: lack of attention can cause falls from scaffolding or from roofs, being hit by swinging beams and hoists, dropping tools and debris, and, for those in casement and caisson work, collapses of retaining walls and other structures.

Construction workers on roofing and paving jobs can experience burns from pitch and bituminous products as well as other industrial chemical exposures. Welders, cutters, and riveters are at risk of burns from heaters, torches, or hot rivets. Those workers involved with high-rise construction are at special risk from falls and from either dropping or having items dropped on them. Welders can forget to lower their shields or to use respiratory protection. Heavy equipment operators can drive backhoes, tractors, or crawlers into ditches or into excavations. Other workers can be run over or hit by swing arms. Crane operators and bucket lifters can drop concrete or equipment onto other workers. Headaches associated with hangovers can lead to reduced attention to details, which can cause accidents. Sedatives and tranquilizers can slow reaction times so that avoiding accidents can be more difficult. Something as simple as forgetting proper lifting techniques can lead to accidents when the worker is impaired by drug use.

The construction industries appear to be less likely to have workers using hard drugs but alcohol is common. Supervisors should remain cautious and observe any behavior or lack of attentiveness on the part of workers; it could indicate drug use.

## ■ 10.6. SUBSTANCE ABUSE IN THE CLERICAL PROFESSIONS

One does not usually associate clerical workers with drug abuse; however, clerks, secretaries, and other office workers may be abusing drugs.

It appears that office workers are more likely to use cocaine in addition to alcohol than those in many other workplaces; perhaps this is due to perceived social status of cocaine use over heroin. Perhaps it is also a result of workers seeking relief from stress often seen in the office, especially one which has many short deadlines and employer/client pressures. Alcohol use on the job seems to be less common than for many other lines of work, most likely due to the potential for detection of the odor of most alcoholic drinks. There is some evidence that office workers who do abuse alcohol either drink vodka or gin rather than whiskeys like bourbon or blended. Use of alcohol as well as other drugs like marijuana and cocaine seems to be more associated with weekend use rather than during the work week.

The employee in the clerical setting who abuses drugs can often cause additional work for others in the office by making mistakes. Work that has to be retyped or re-filed or that is even lost can lead to a decrease in productivity for an office. A fairly common situation in many offices is the cover up of one worker's abuse by co-workers; often this results from simple friendship or a desire to keep the office productivity high and problems away from the supervisor. Workers who abuse drugs in this setting are ones who often take off early on Fridays and suffer from decreased productivity on Mondays. Such workers often make mistakes that have to be corrected by others and experience mood swings or evidence of paranoia. Long lunch hours may be another symptom of a drug problem in the business office.

## ■ 10.7. SUBSTANCE ABUSE IN THE LAW ENFORCEMENT AND FIREFIGHTING PROFESSIONS

Two of the most stressful of jobs are law enforcement and firefighting. Stress makes police work a likely candidate for misuse of chemical substances as does almost daily contact with criminals and others often connected with the chemical substance distribution networks. The boredom of waiting for a fire or accident combined with the stress of actually fighting a fire or containing a hazardous materials accident makes firefighting almost as stressful. It is unfortunate and tragic if a policeman or fireman goes wrong. Police, who are depended upon to protect society from itself, who fail can cause more harm than just to themselves since the fabric of society is torn; fire protection personnel are likewise looked upon as always being ready to risk their lives to protect the property or lives of others. Luckily, chemical substance abuse is rare among law enforcement and firefighting personnel, largely due to well-organized and extensive drug testing and abuse identification programs. Sometimes, however, the availability of chemical substances due to contacts with dealers, manufacturers, prostitutes, and others involved with chemical substances creates a temptation too strong to resist and a few enforcement personnel fall into abuse. The life of a firefighter alternates between boredom while waiting for a call, to life-threatening situations involving destruction and death; such conditions can lead to chemical substance misuse. Since most public protection services have chemical substance testing programs, abusers are usually caught in a timely manner and rehabilitated or removed from service.

The most common substances abused seem to be beer and other alcoholic drinks in the law enforcement and firefighting communities; other chemical substances are rare and the "hard chemical substances" extremely rare. Because of work stresses and even traditions, many law enforcement and fire personnel socialize after work by having a few drinks with fellow officers. While most such personnel do not drink to excess, a few do. This would not be a problem for most workers, but, since police and firefighters are required to be available at all times, emergencies can happen while the individual is not in top shape.

Although chemical substance abuse is rare among law enforcement and firefighting personnel, the consequences are potentially very serious since almost no other professions require employees to be so mentally alert at all times. Criminal situations and accidents can occur without warning at any moment; enforcement and protection personnel either on duty or off duty, must remain alert and aware of conditions around them. There is evidence that times are changing and that police and fire personnel are becoming more aware of the dangers of alcohol through active programs of the various public service agencies.

The most serious problem for law enforcement and firefighting personnel with respect to chemical substance abuse remains the misuse of alcohol. It is hoped that self control by police and firefighters will reduce or even eliminate the problem which exists.

## ■ 10.8. SUBSTANCE ABUSE IN THE MILITARY

The military received a lot of unfavorable criticism during the fighting in Southeast Asia during the 1960s. Most of the criticism was unfortunately deserved, because there was widespread misuse of chemical substances in that theater of operations. Marijuana and alcoholic drinks were abundant and even expected to be used by many, if not most, of the soldiers. Even more dangerous was the easy availability of "hard chemical substances", especially heroin, due to the proximity of many of the illicit chemical substance trade routes and production areas. The costs of chemical substances were very low, so, coupled with the stresses of war, death, and knowledge that many at home were violently opposed to the war, it created an environment in which chemical substance abuse was considered desirable and was quite common. Military programs to reduce or eliminate the problem were generally not effective for a variety of reasons. There is some evidence to support the argument that chemical substance abuse increased as domestic opposition to the war escalated and as the fighting dragged on without any clear evidence of winning.

But, that is the past! Following the end of fighting, the military successfully took bold and decisive steps to stamp out chemical substance abuse. Urinalysis testing without notice became common for soldiers, sailors, marines, and airmen from enlistees to generals. Regulations were written and implemented to attempt to rehabilitate offenders and then to swiftly remove them from service if such attempts failed. Penalties for officers were especially strict: separation from the service upon one positive test result. Improved respect for the military by the general public may also have played a role.

However, this is not to say that the military is without chemical substance abuse, especially since the use of alcohol is still common, although recent actions such as forbidding "happy hours" at military clubs, warning labels on alcoholic drinks, and command policies curtailing alcohol use at official or even unofficial functions have tended to somewhat reduce the use of alcohol. The military services have programs to identify and to rehabilitate those who misuse chemical substances. The alcohol program has been quite successful. The use of hard chemical substances is today fairly rare in the military due to mandatory chemical substance testing programs for personnel of all ranks and organized programs at every installation to identify abusers.

The biggest chemical substance abuse problem in the military remains alcohol, since it is a lawful substance and available at low prices in clubs and packages stores (Class VI stores, in military parlance). Marijuana use is much less common but is sometimes identified during chemical substance testing. The use of other chemical substances is relatively rare in the military; however, misuse of over-the-counter or prescription chemical substances is largely unknown (as it is in other sectors as well) since the chemical substance testing is aimed at opiates, amphetamines, marijuana, and alcohol.

The days of extensive chemical substance abuse in the military have largely passed and the public can rest assured that the demoralizing and efficiency-destroying conditions of the Vietnam War period are gone; the problems which remain are being handled by the services effectively and quickly.

■ 10.9. SUBSTANCE ABUSE IN THE ENTERTAINMENT INDUSTRIES

Hollywood is well-known for its use of drugs, even in the early years. Famous actors and actresses have received much publicity about overuse of alcohol and other chemicals. The pressures of public life as well

as extremely large salaries at the top of these fields lead to easy availability and perceived need for drugs.

Although most performers do not earn the fantastic salaries the public
often associates with entertainment, there is a perceived environment
in which such people have to take part in parties at which alcohol and
other drugs, especially cocaine, are common. In few other workplaces
is there such a need to be recognized and be a part of the accepted
crowd. Those who refuse to take part will not be trusted and fail to
win bookings and parts. In addition, most public performers entertain
in bars and lounges, at least at some time in their career; the availability and the environment there lead to consumption of alcohol and, often, other drugs. Entertainment workplaces, from lounges to radio studios to television studios to movies, are often very stressful, due to
limited availability of good roles or gigs and the fact that most produc-
ers are very close knit, hiring only those with a proven track record of
bringing in customers. Pressures associated with deadlines and keeping to a schedule increase the "need" for stimulants to keep up the
pace, followed by tranquilizers to "wind down".

The most common drugs used in the entertainment industries are
alcohol, cocaine, stimulants, and tranquilizers. Although some enter-
tainers, no doubt, are hooked on heroin, it does not seem to be as
common as in other lines of work.

While these lines of work often receive much notoriety, especially
when very famous people are involved, the use of drugs is not limited
to these workplaces; almost any office, job site, or factory has some
drug abusers.

## ■ 10.10. SUBSTANCE ABUSE IN SPORTS

While news reports of sports figures found to have been abusing drugs
are headline affairs, the number of athletes abusing drugs is probably
small. Most athletic performers are very concerned and aware of their
bodies. Although a few noted sports stars have been involved in drug
usage, it is most likely a result of wealth rather than sports. In fact,
some of the cocaine-related deaths seem to have resulted from first-
time use of the drug by the person. Even the use of alcohol by most
sports figures is much less than commonly suspected, due to the requirements to be able to perform at their peak all the time. The most
commonly abused substances for sports figures seem to be dietary supple-
ments and steroids, although recent revelations about the harm from
steroids and routine testing of star athletes have reduced the usage
from a few years ago.

Many noted sports figures have taken part in education programs
designed to warn young people about the dangers of drugs; these have

been quite successful since the kids respond to their heroes more than to teachers or even parents.

Most professional sports figures understand that abuse of drugs can mean a decrease in performance; with the highly competitive nature of sports today, they can't afford to be performing at less than their best. Minor league players, lesser-known figures, and the aspiring young athlete might be tempted to try drugs in order to enhance (or appear to enhance) their performance. Generally, these will never make it to the top of their profession.

The few noted figures who are suspended, fired, or fined for drug abuse seem to make more of an impression on others than on themselves; there are often repeat offenders, although some heed the warning and stay off the drugs which can be a detriment to their careers. The reputation for drug abuse in professional sports seems to be somewhat overblown; however, since these athletes are heroes to young people, even one case can be a serious matter.

## ■■ 10.11. SUBSTANCE ABUSE IN THE TEACHING PROFESSION

A teacher who abuses drugs is a tragedy, since teachers are the mentors of the next generation. It is very unfortunate when a teacher becomes involved in drug misuse since, even today with highly cynical students, teachers are still looked up to for guidance. Some school boards have instituted mandatory drug testing, but these programs have largely been met with resistance from teachers' unions. Such testing is considered an insult to the dedicated teachers and an infringement on their rights.

Few teachers involve their students in drug abuse; teachers remain a group largely opposed to the use of drugs and dedicated to sincerely helping young people not get involved. In fact, there have been very few cases of teachers pushing drugs or involving students in drug use. Most secondary and even some elementary school teachers today have seen students high on drugs in their classes; many college professors have, also. Since the majority of teachers are sincerely involved in helping young people, they usually try to turn a weak student from drug abuse.

Teachers, like most Americans involved in drug misuse, are sometimes involved with alcohol and even fewer with cocaine. Because of the profession, most teachers avoid public displays of improper behavior. As a result of pressures, there are a few teachers who abuse tranquilizers; some also are involved with stimulants in order to pep up. Even rarer is the professor who invites students to join in marijuana use. Drug abuse does not appear to be a serious or widespread problem for the teaching community.

■■ 10.12. SUBSTANCE ABUSE IN
THE MANUFACTURING AND
UTILITIES INDUSTRIES

The so-called "blue-collar" workers in the manufacturing and utilities industries most commonly abuse alcohol. The use of marijuana appears to be second but far back in popularity. Use of cocaine is not very common among these groups. Only rarely, heroin and other drugs are abused.

Because of similar likes and dislikes, many factory workers will congregate at a bar on the way home from work or on weekends. Beer and malt liquor are most commonly consumed during socialization at bars or watching television. If involved with drugs at all, the "typical" manufacturing worker abuses alcohol, most likely in a social setting.

It appears that misuse of tranquilizers is quite rare among factory workers; there is some evidence that some workers, especially those on "midnight" or "graveyard" shifts, are involved with use of stimulants, probably to help stay alert and awake during night-time hours.

The abuse of drugs by these workers is especially hazardous since many work with dangerous materials or in dangerous situations. Alertness is important in order to prevent accidents; unfortunately, the use of stimulants may give the impression of alertness, while fatigue of the mind and body continue, with the result that mistakes can be made. Many industrial accidents occur because of worker inattention; drug misuse can increase reaction times and even decrease awareness of the seriousness of the situation. Workers in manufacturing and utilities should be among the most careful with respect to misuse of drugs since their work is often dangerous or at least potentially so.

■■ 10.13. SUBSTANCE ABUSE IN THE WAREHOUSING
AND SALES INDUSTRIES

Workers in warehousing and sales, like the factory workers above, are most commonly involved with alcohol, often in a social setting. Marijuana use is probably second, followed by misuse of stimulants.

Warehouse workers often deal with heavy equipment and heavy containers, so accidents can be very serious. These workers should be very careful of using any drugs which reduce alertness or perception, even over-the-counter antihistamines and similar medicines. Mixing alcohol and medicines can often increase the danger of sleepiness.

Sales people often work under pressure of deadlines and quotas; they can be involved with stimulant misuse. Misuse of tranquilizers seems to sometimes occur, especially after long work hours, stressful sessions, or time on the road. Sales people, especially outside ones, are generally aggressive and out-going people by nature, and so often are involved in social situations at which drinking is expected. A danger exists of becoming used to overdoing alcohol.

For both of these groups, the most commonly abused drug is alcohol.

## ▬ 10.14. SUMMARY

People from any profession or job can become addicted to chemical substances with potential danger to themselves and others. The misuse of drugs can cause accidents, incorrect administration of medicines, inattention to details, disruptive behavior, slips and falls, and numerous other dangers. Chemical dependency cuts across all socioeconomic classes, professions, and ages.

The important thing to remember is that the drug abuser needs help, possibly for more than the drug habit, since the misuse of drugs by workers often results from something missing in the person's life or from real or perceived stress.

# 11 CASE STUDIES AND EXAMPLES

## 11.1. USE OF CASE STUDIES

Although each case of drug misuse is different, since different people are involved, there is much to be gained from looking at cases in which drug misuse led to accidents, injuries, and death. The following case studies are from the files of the U.S. Department of Labor's Occupational Safety and Health Administration (OSHA) and Mine Safety and Health Administration (MSHA). The authors would like to thank Joseph DuBois, Ph.D., OSHA, and Frank Swamberger, MSHA, for their assistance in obtaining the following information.

## 11.2. CASE STUDIES

The following 30 cases are examples of when injuries or death may result from improper use of drugs, including alcohol and prescription medications.

**Case Number 1**
An example of when drug use can lead to death in an occupational setting when the employees were not at fault is an accident which occurred when two employees, aged 49 and the 29, were cleaning and patching expansion joints between concrete slabs on a highway. The driver of a stolen car, under the influence of drugs, swerved into the closed lane and struck the two workers. The 49-year old was killed and the 29-year old severely injured.

**Case Number 2**
A case in which a 29-year-old worker caused his own death as a result of alcohol consumption occurred when an employee was separating copper, brass, and aluminum scrap as it passed in front of him on a

conveyor belt. Unfortunately, he was working alone. He was dragged by the conveyer into a wall, where he was found by fellow workers. He died due to crushing injuries with extensive lacerations with his right elbow dislocated and fractured. He had been drinking alcoholic beverages just prior to his death.

### Case Number 3
An interesting case in which death resulted from climatic exposure and alcohol misuse was that in which a 52-year-old employee was to watch logging equipment at night during a period of "shut down". A relief crew found him dead and notified the Sheriff's Department. The county Medical Examiner found the cause of death to be hypothermia due to acute alcohol intoxication.

### Case Number 4
An example of a fatality resulting from impaired judgment occurred when an employee was assigned to open hatches and pump water out of a barge in preparation for taking it out of the water for drydocking. The 28-year-old employee, under the influence of drugs, removed a sealed hatch of a double-bottom tank and entered it before it had been tested for oxygen deficiency. In fact, a later test showed the oxygen level to have been as low as 8% (21% is normal and <15% dangerous). His body was found later by co-workers.

### Case Number 5
A less serious accident, yet one which did lead to injury, was one in which a 25-year-old employee was feeding a hose into a rod hole. The operator's hand slipped, allowing his hand to enter the rod hole, resulting in amputation of his little finger. The employee tested positive for controlled substances.

### Case Number 6
Another injury resulted when a drug-impaired 23-year-old employee received a fractured arm due to a falling shackle while working on pulling polished rod. The worker's record had never indicated any previous accidents or evidence of drug abuse. The victim tested positive for several drugs.

### Case Number 7
Another death resulted when an employee failed to be alert, due to alcohol impairment. The employee (#1) was directing a fellow employee (#2) to back up a Mack truck to a waiting asphalt-paving machine. While employee #2, the driver, was backing the truck, employee #1 fell down out of view of the driver. Employee #1 was crushed to death. The deceased 66-year-old employee was found to have in his posses-

sion a bottle of liquor, indicating drinking while at work. Autopsy results produced a blood alcohol level of 0.112%.

## Case Number 8
A case in which a health care worker abused medications occurred when a 39-year-old employee was performing home health care on a patient. The employee administered a dose of morphine to herself, which resulted in a drug overdose and subsequent fatality. The morphine had been apparently taken from the patient's medication, as 4 tablets of MS Contin (14 mg) and 24.5 cc of Roxinof (liquid), which are morphine products, were missing from the patient's inventory.

## Case Number 9
An accident which led to death involving alcohol abuse together with other drugs occurred when a 35-year-old employee fell into a coal bin and suffocated. Acute ethanol intoxication in combination with narcotic drugs was listed as the cause of death.

## Case Number 10
Another example of an employee, under the influence of alcohol, causing his own death, was when two employees were installing a highway sign. They faced each other and alternated sledge hammer swings at the signís stub. Employee #1 hit a glancing blow on his co-worker's hammer while it was on the stub. A metal shard dislodged from his hammer, flew back, and punctured his chest cavity. The 23-year old employee bled to death from his wound; he had a blood alcohol level of 0.068%.

## Case Number 11
A case in which an impaired driver caused the death of another person occurred when a residential refuse truck was pulling away from the curb after picking up trash. A van struck the truck from the rear, knocking a 40-year-old refuse employee into the truck and then into the street. The victim died instantly of his injuries. The van driver, subsequently charged with second degree murder for driving under the influence of alcohol, was also injured in the accident.

## Case Number 12
An incident in which two workers were killed by another under the influence of legal but excessive drug usage took place when two laborers walking along the paved side shoulder of a road were struck by an automobile going in the same direction. One laborer, aged 34, died instantly of massive head injuries, while the second, aged 31, died in the hospital later. The driver of the car was charged with involuntary manslaughter when prescription drugs were found in the automobile and his blood.

### Case Number 13

Another accident resulting from impaired behavior occurred when a 48-year-old employee failed to complete proper shutdown procedures before opening the door to a 200 foot water tower prior to taking samples. The gate that should have shut off the water had not been energized, and the employee had not checked the chamber to see if it was empty. When he opened the door, he was struck by 20,000 gallons of water. The medical examiner's autopsy indicated high levels of prescription and contraband drugs in the victim's system.

### Case Number 14

Another case of a medical worker abusing hospital supplies occurred when a nurse was found slumped over a cylinder of nitrous oxide in the nurses' lounge. The employee was working alone to clean the operating rooms where nitrous oxide cylinders were kept. There was a broken seal on the nitrous oxide cylinder and a "T" handle wrench used to open cylinder valve was found at the victim's feet. Medical examiner's report indicated high levels of nitrous oxide, but no alcohol or other drugs.

### Case Number 15

Employee #1, the 40-year-old victim, was a janitor assigned to clean and mop in the theater area of a large metropolitan civic center. It was common practice for janitors to open elevator doors from the outside of the landing by inserting a scraper between the doors and prying the doors open. Apparently, the victim had opened the third-floor doors with the elevator present and then walked away. Another janitor noted the open elevator and closed the doors, after which the elevator was called to another floor. Upon returning, employee #1 opened the door using his scraper, failed, due to alcohol impairment to note that the car was no longer there, and fell into the elevator shaft to the top of the car on the next floor down. He died as a result of his fall.

### Case Number 16

Smoking in bed is always dangerous, but when alcohol intoxication is added, the combination can be fatal. A 33-year-old employee with a blood alcohol level of 0.28% fell asleep in a rest trailer at work. He was resting on a cot while smoking; the trailer caught on fire. Unfortunately, the trailer did not have smoke detectors installed. The employee was overcome by smoke and carbon monoxide and was found dead at the scene.

### Case Number 17

A common accident occurred when a 34-year-old alcohol-impaired employee was repairing the roof of a dwelling, slipped, and fell to the ground, dying as a result of his injuries. Unfortunately, in addition to the drug abuse, adequate safety devices had not been installed.

**Case Number 18**
A case involving marijuana use in the workplace took place when a 27-year-old employee was working along a sloping, shingled house roof with an eave height of approximately 16 feet and a slope of 4 in 12. At the time of the accident, he was apparently walking from the actual work location on the roof to the access ladder when he slipped and went off the roof. His head struck the ground below, causing critical injuries. He died 2 days later. There were no witnesses to the actual event, but a drug screen indicated that the employee was positive for marijuana.

**Case Number 19**
Another case of impaired judgment resulting from alcohol use on the job together with improper supervision occurred when a 52-year-old warehouse employee had been directed to go home by the company medical department and his foreman because he smelled of alcohol and evidenced drunk behavior. Unknown to his supervisors, the employee sneaked back into the warehouse and went to sleep in a remote corner where he was eventually found dead. It was found that he had died from exposure to toxic substances because he was found on the site of a paint spray operation that used automobile acrylic enamel.

**Case Number 20**
Even farm workers can suffer injury and death from alcohol use on the job. A 27-year-old farm worker was performing irrigator work. He was seen driving a 4-wheel all terrain vehicle (ATV) at a high rate of speed over a dirt road along the side of the irrigation ditch. Fellow co-workers later found the ATV's headlights pointing up to the sky and the ATV upside down on the injured employee. He died from massive head injuries. Alcohol was the cause of his inattention to the hazards of drving.

**Case Number 21**
Combining alcohol with industrial chemicals can be fatal. A 30-year-old laboratory technician reported for work on the night shift of a Texas cement plant. It was later discovered that, at that time, his blood alcohol content was at the intoxication level. Unfortunately, he was allowed to go to work. Some time later, apparently to add to his "high", the man connected a makeshift mask and hose to nitrogen piping and began breathing near full-strength gas. The technician, who had 6 years' mining experience including 4 years with the analytical laboratory, apparently was unaware of the gas's ability to displace oxygen. He was found dead in the laboratory 1 1/2 hours later. An autopsy showed that he died of suffocation.

## Case Number 22

A 42-year-old truck driver, who had more than 6 years' of mining experience, died when he backed his haulage truck over a rock dump at a Minnesota taconite mine under inadequate lighting conditions. The accident occurred as the victim, who had failed to fasten his seat belt, drove too close to the stockpile edge, causing the bank to collapse and the truck to roll over. The truck's brakes were found to be in good order, but there were no signs that the driver had ever applied them while on the stockpile. Autopsy results revealed that the driver had been in a state of acute alcoholic intoxication.

## Case Number 23

A 45-year-old truck driver with 25 years of experience was crushed to death near a surface coal stockpile in West Virginia when he inadvertently backed an endloader across a roadway berm, and the machine plunged down a steep enbankment and overturned. The victim, who was not a regular endloader operator, was found by a second shift foreman, pinned under the left rear wheel. The endloader was found to be in good operating condition. According to the foreman, the victim had been discharged for being intoxicated on duty 9 months before the accident. A bottle found in the victim's truck contained 16% alcohol, and the autopsy revealed that the victim had been under the influence of alcohol when he attempted to operate the equipment.

## Case Number 24

A truck-mounted highwall drilling machine tipped over an enbankment near a haulage road at a West Virginia surface coal mine. The 32-year-old operator, who had less than 6 1/2 months' mining experience, was found dead, his body wedged between the frame of the operator's compartment and the control panel. An autopsy showed that the victim's blood alcohol content was close to the legally defined intoxication level when he died. An investigation showed that there had been no deficiencies in the drilling machine, which was found to have traveled about 30 feet on the berm before going over the bank, so alcohol appeared to be the sole cause.

## Case Number 25

Luckily not all accidents lead to death although injuries may result. A road grader operator, aged 54, at a coal mine in Illinois was injured when he ran off the road, down an embankment, and into a drainage ditch. He was thrown through the windshield as the grader plunged into the water. After the accident the victim was pulled from the water, and a check of the grader showed that there were empty beer cans on the floor of the cab and additional unopened cans in a nearby cooler. No brake or mechanical defects were found in the inspection of the grader. The employee was said to have had a drinking problem in the past, but apparently had not had any recent problems with alcohol until one month prior to the accident. A blood sample taken after the accident showed that the man had a blood alcohol level nearly three times the legal definition for intoxication. He later entered a treatment program for alcohol problems.

## Case Number 26
A 48-year-old tractor operator was killed after driving to work late at night in his personal vehicle when he crashed into a disabled haulage truck on a mine road that was supposed to be used only by vehicles of a midwestern coal company where the man was employed. The driver of the haulage truck had warned each approaching vehicle to slow down by waving his lighted flashlight after the headlights on his truck had failed. The tractor operator apparently had not seen the parked haulage truck or the waving flashlight in time to prevent the collision. An autopsy disclosed that the tractor operator was intoxicated at the time of death.

## Case Number 27
A 22-year-old laborer was killed when she fell 74 feet through a cleanout hole in a work deck at the top of a silo under construction to the floor. She had less than 3 months of experience in mining. Two marijuana cigarettes were found in the victim's hardhat. A co-worker reported seeing the victim smoking "pot" prior to the accident. A foreman had observed that several persons had been smoking marijuana that day prior to the accident, and company officials said they had reprimanded workers for using drugs in the past but apparently without affecting either the workers' behavior or restricting their employment.

## Case Number 28
A 44-year-old front-end loader operator for a sand company in the Southeast, with 1 year and 9 months of experience at his job, was suffocated after he left his cab at a sand pit and walked to a point near the right front of the loader, where part of a highwall collapsed on him. He was found buried in 3 to 6 feet of sand. The loader was equipped with cab rollover protection and seat belts. A glass jar containing alcohol and seven empty beer cans were found in a cooler inside the cab of the loader. Although cause of death was attributed to suffocation, the autopsy report showed that the victim's blood alcohol content was more than twice the legally accepted intoxication level.

## Case Number 29
A 64-year-old truck driver for a midwestern quarry was drowned when he backed his haul truck over into the quarry, causing the truck to turn upside down in the water. The victim, who had 16 years of mining experience, apparently had intended to dump a load of waste over the

bank rather than at the site he was instructed to use. An autopsy disclosed that his blood alcohol was twice the legal level.

**Case Number 30**

A 30-year old dayshift equipment serviceman at a surface coal mine was injured when he tried to sit on a mound of dirt near a highwall crest and apparently slid or fell over the bank. For unknown reasons, the man had wandered from his designated assignment at a drill rig to the highwall area nearly a quarter mile away. The man, who had about 7 years' mining experience, had been suspected of drinking on the job by another employee who smelled alcohol on his breath that morning. An empty beer bottle was later found in a bulldozer cab where he had been working, and a capped pint bottle about one-third full of 190-proof grain alcohol was found beneath the seat of a service truck the victim had driven that day.

## ■ 11.3. SOME GENERAL CONCLUSIONS FROM THESE CASE STUDIES

A common thread connecting these and other cases studies is that the most common drug of abuse in the workplace is alcohol. Accidents and death may result to the abuser or to other workers or even innocent by-standers. Supervisors have a responsibility to their workers and to the company to ensure that impaired workers are not allowed to injure themselves or others. This requires that supervisors evaluate each employee regularly and ensure that when workers are sent home or to treatment they actually do so. These case studies are some of the more obvious and fatal of incidents; it is impossible to determine the impact of reduced employee performance and slipshod quality on a company's well being. Drugs are being abused by workers on and off the job site or office; supervisors and management need to be aware of the dangers of injury and even lawsuits which can result from employee drug abuse.

# 12 A FINAL WORD

## 12.1. WHY THIS BOOK

As we began this book, we gave the reasons for writing it: a sincere concern about the dangers of the drug abuse epidemic taking its toll on health and productivity and even the spirit of the nation. Drug abuse is a serious matter in our society, harming families, workers, and society. The drug underculture reeks with crime and criminal activity. Fortunes are wasted and drug lords become rich. Even the security of our nation is weakened by the flow of money from our borders, decreased productivity, wasted lives, hopelessness, and a cycle of dependency, often involving other crimes to pay for the habit of the addict. The destruction of the minds of the young and the bodies of the old continues and increases. Workplaces are no longer places where people strive together to produce the best possible product, whether it be a piece of furniture, a building, efficient service, an enjoyable play or song, food from the farm, or the wealth of the earth and seas. Federal law, in the Drug-Free Workplace Act of 1988, has mandated that employers and employees together work to reduce the decrease in productivity and harm to society caused by improper use of drugs.

## 12.2. WHAT CAN BE DONE

Chemical substance abuse has been with us since the beginning of time and will most probably continue well into the future. So, why should we try to stem the tide? The effects on society and the economy are staggering; lost time and decreased productivity due to impairment, injuries, and accidents amount to billions of dollars annually. The ruin of human lives is an even greater loss. Education is vital to bringing about an understanding of the harm from drug abuse, but there is more. Much of the abuse of drugs has deeper social causes: poverty, free time without constructive activities, high drug profits with relatively low risk, frustration and anger over the failure of society to take care of the poor and homeless, and a move away from family values and stable home environments. To remove the specter of drug abuse from society, all these causes have to be addressed. This is not to reduce the importance of individual responsibility; each drug user seeks to find personal pleasure or escape from reality in an egocentric environ. All of us need to work to remove the causes of injustice and poverty, tension and hate, and inhumanity.

Helping drug abusers is more important than punishing them. If there is not demand for drugs, there will be no crime involvement and no supply. We must all work to help those who feel they need to hide inside a bottle or vial or syringe. Until the day comes when there is no poverty and hopelessness, we have to work to educate and help the abuser.

Use of all chemicals should be in moderation and only when absolutely needed. Use of alcohol should be moderate. The taking of medications, whether prescription or over-the-counter, should be limited to the quantity and duration needed to accomplish return to health.

## ■ 12.3. CONCLUSION

The epidemic of drug abuse can be ended. Help to those who need it must be available in EAPs, in poverty-prevention programs, in elimination of racial and religious hatred, and in general society. The American workplace can be made drug-free if everyone involved, unions, employers, workers, and the public, want it to be. Education remains the most important weapon in this war. Best wishes in making your workplace as safe, healthful, and drug-free as possible.

# APPENDIX A

## CHRONOLOGY OF HISTORY OF DRUG ABUSE

| | |
|---|---|
| 5000 B.C. | Earliest recorded record of opium use by Sumarians. |
| 1500 B.C. | Chichimec tribe in Mexico used mescaline in their religious ceremonies. Teonanacatl "food of the gods", the magic mushroom (psilocybin), was used in ceremonies. |
| 300 B.C. | Greek physician Erasistratus warns of the addictive properties of opium. |
| 1000 A.D. | Beginning of coca cultivation in Peru. |
| 1200 | Opium introduced into China by Arab traders. |
| 1208 | The coca plant is declared a gift of the Sun God in the Incan culture; it takes its place as a divine plant. |
| 1565 | Dr. Nicholas Monrodes publishes first description of coca in Europe. He praises the plant in combating hunger and fatigue. |
| 1729 | China prohibits the sale and smoking of opium. |
| 1762 | "Dover's Powder", containing opiates, is introduced in England and becomes one of the most common medical preparations for the next 150 years. |
| 1805 | Morphine is isolated and described by Friedrich Wilhelm Adam Serturner, a German chemist. |
| 1839–1842 | The First Opium War. The British forced the trade of opium on China, a trade the Chinese wanted to prohibit. |
| 1848 | National Drug Import Law passed in U.S. |
| 1856 | The Second Opium War. |
| 1859 | Paolo Mantegazza, an Italian physician, proclaims that coca leaves are a great new weapon against disease. |
| 1860 | Albert Niemann isolates an alkaloid from the coca leaves and calls it cocaine. |
| 1861–1865 | 45,000 soldiers become addicted to morphine during the American Civil War. |
| 1864 | A Dr. Nusbaum in the U.S. is the first practitioner to mention increased dangers of abuse of hypodermic injection of morphine. |
| 1874 | Heroin was first discovered by C.R. Alder Wright, a London chemist. |
| 1880 | Importation of opium is banned in China and the death penalty is imposed for its use. |
| 1880 | Cocaine is used in the treatment of morphine addiction. |

| | |
|---|---|
| 1884 | Sigmund Freud takes cocaine for the first time. |
| 1884 | Freud and Bruer treat their friend and colleague, Von Fleischl's, morphine addiction with cocaine. |
| 1885 | An Iowa survey finds 3000 stores which sell opiates across the counter without prescription. |
| 1885 | Robert Louis Stevenson, while under the treatment with cocaine for tuberculosis, wrote the first draft of Dr. Jekyll and Mr. Hyde, in 3 days. |
| 1885 | Park, Davis & Company manufactures 15 preparations of coca and cocaine including Wine of Coca and Coca Cigarettes. |
| 1887 | Opium Act passed, forbidding import of opium into U.S. by Chinese, and prohibiting U.S. citizens from the opium trade in China. |
| 1888 | Coca Cola, introduced by Asa G. Chandler, contains cocaine and coca oil extract and is advertised to "cure your headache" and "relieve fatigue" for only $0.05. (Cocaine and coca oil were later removed.) |
| 1888 | The actions of peyote and mescaline are described by Louis Lewin, a German pharmacologist. |
| 1896 | The active ingredient (mescaline) of the cactus *Lophophora williamsonii* is isolated. |
| 1906 | U.S. Pure Food and Drug Act requires patent medicine labels to list all "dangerous substances" contained in the mixture. |
| 1909 | Opium Exclusion Act prohibits import or use of opium except for medicinal purposes. |
| 1912 | The Hague Conference agreement reached that production and trade of opiates and opium be limited to the amounts necessary for medical and scientific use only. |
| 1914 | Harrison Narcotic Act (a U.S. tax law): persons authorized to handle or manufacture drugs required to register, pay fee, and keep records of all narcotics in their possession. |
| 1919–1924 | Establishment of Public Outpatient Narcotic Clinics in hope of rehabilitating the addict and preventing his involvement with criminal drug distributors. In general, badly managed; by 1924, all were forced to close by a moralizing and crusading press and the Federal Bureau of Narcotics. Illicit narcotics became the addict's only source of supply. |
| 1920 | Volsted Act  Volstead prohibits nonmedical usage of alcoholic beverages. |
| 1922 | Behrman Case decision prevented medical doctors from legally supplying drugs to addicts for self administration and implied that addicts must be isolated and hospitalized. Led eventually to the Public Health Service hospitals in Lexington, Kentucky (1935) and Fort Worth, Texas (1938). |

| | |
|---|---|
| 1922 | Jones-Miller Act (U.S. Narcotic Drug Import and Export Act) established firm penalties for violation of the Harrison Act. |
| 1924 | Prohibition of manufacture of heroin in the U.S. |
| 1925 | Supreme Court ruled that a physician may administer narcotics to allay withdrawal symptoms if done in good faith. The Federal Bureau of Narcotics ignored this ruling, punishing physicians who gave narcotics to addicts. |
| 1930s | MDA (Mellow Drug of America, Love Drug) synthesized. Very similar to amphetamine and mescaline. |
| 1933 | Repeal of Prohibition of alcohol in the U.S.. |
| 1937 | U.S. Marihuana Tax Act brings marihuana under stern controls similar to those regulating the use of opiates. |
| 1938 | Food, Drug, and Cosmetic Act regulated quality and purity of food, drugs, and cosmetics sold in interstate commerce in the U.S. |
| 1942 | Opium Poppy Control Act required registration and taxation of growers of opium poppies in the U.S. |
| 1951 | U.S. Boggs Act graduated sentences with mandatory minimum sentences for all narcotic drug offenses. Subsequent to the passage of the Boggs Act, many State legislatures enacted "little Boggs acts." |
| 1956 | U.S. Narcotic Drug Control Act was even more punitive than the Boggs Act; it did, however, differentiate among drug possession, drug sale, and drug sales to minors. Medical use of heroin was prohibited. |
| 1956 | All existing heroin supplies in the U.S. were surrendered to the government. |
| 1960s | Use of psilocybin, LSD, and other psychedelics gain in popularity. |
| 1963 | Supreme Court (Robinson vs. California): Declared that addiction is a disease, not a crime. Legally, an addict cannot be arrested for being "high" (internal possession) but he can be arrested for the external possession and sale of drugs. |
| 1966 | Narcotic Addict Rehabilitation Act views narcotic addiction as being symptomatic of a treatable disease and not a criminal condition. |
| 1966 | Drug Abuse Control Amendments became effective, whereby sedatives and stimulants came under tighter controls. Hallucinogens were specifically added to the laws in 1966. Enforcement became the responsibility of the Bureau of Drug Abuse Control in the Food and Drug Administration (FDA). |
| 1968 | Bureau of Narcotics and Dangerous Drugs (Department of Justice) was given responsibility on a federal level for the entire drug problem. The Bureau of Narcotics was removed from the Department of the Treasury and the Bureau of Drug Abuse Control from the FDA, and the two were combined. |

| | |
|---|---|
| 1969 | Operation Intercept: An attempt to block import of marijuana at the Mexican border. It coincided with increased use of "harder" drugs throughout the country. |
| 1969–1971 | Tightening of controls at a federal level and urging of foreign governments to apply firmer restrictions in regard to manufacture and exportation of drugs. |
| 1970 | U.S. Comprehensive Drug Abuse Prevention and Control Act of 1970 replaced previous acts for control of narcotics, marijuana, sedatives, and stimulants; and placed their control under the Department of Justice. Drugs are classified into five schedules according to their potential for abuse and therapeutic usefulness. First-time illegitimate possession of any drug in the five schedules is considered a misdemeanor and penalties are reduced. Provisions are made for rehabilitation, education, and research. House search ("no-knock" law) legalized. |
| 1972 | Drug Abuse Office and Treatment Act brought about by the increasing drug use by U.S. troops in Vietnam and increased use in the U.S., this law established the Special Action Office for Drug Abuse Prevention (SAODAP) to be the coordinator of the nine Federal agencies involved in drug activities. With an emphasis on treatment and rehabilitation programs, SAODAP develops Federal strategies for all drug abuse efforts outside of drug traffic prevention. Also detailed in the legislation is the establishment of the National Institute on Drug Abuse, which took place in April 1974. This organization continues the programs established by SAODAP. |
| 1988 | Passage of the Drug-Free Workplace Act which requires that most employers establish policies and programs to educate and help workers avoid drug abuse. |
| 1990s | Some reduction in drug abuse as education programs implemented. Crack cocaine becomes drug of choice in poorer neighborhoods. Alcohol free beers introduced. LSD makes slight comeback. |

# APPENDIX B

## CONTACTS AND TELEPHONE NUMBERS FOR HELP

**For assistance with drug abuse concerns and drug-free workplace information:**

The Department of Transportation (DOT) Anti-Drug Information Center (800-CAL-DRUG).

The Employee Assistance Professionals Association (703-522-6272).

The Institute for a Drug-Free Workplace (202-842-7400).

The National Association of State Alcohol and Drug Abuse Directors (202-783-6686).

The National Clearinghouse for Alcohol and Drug Information (800-729-6686).

The National Institute on Drug Abuse Treatment Hotline (800-662-HELP).

**For information on listing of laboratories or the National Laboratory Certification Program:**

Department of Health and Human Services (301-443-6014).

**For state attorneys general and state alcohol and drug abuse officials:**

See pages 146 and 147 in Appendix D.

# APPENDIX C

## PROCEDURES FOR SETTING UP DRUG TESTING PROGRAM
### (Courtesy of the National Institute on Drug Abuse)

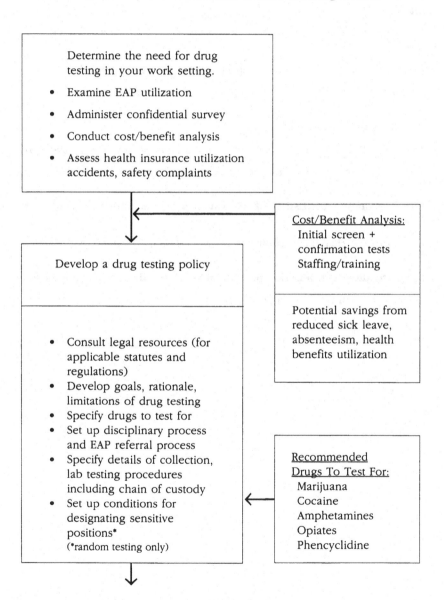

Determine the need for drug testing in your work setting.

- Examine EAP utilization
- Administer confidential survey
- Conduct cost/benefit analysis
- Assess health insurance utilization accidents, safety complaints

Develop a drug testing policy

- Consult legal resources (for applicable statutes and regulations)
- Develop goals, rationale, limitations of drug testing
- Specify drugs to test for
- Set up disciplinary process and EAP referral process
- Specify details of collection, lab testing procedures including chain of custody
- Set up conditions for designating sensitive positions*
  (*random testing only)

Cost/Benefit Analysis:
Initial screen + confirmation tests
Staffing/training

Potential savings from reduced sick leave, absenteeism, health benefits utilization

Recommended
Drugs To Test For:
Marijuana
Cocaine
Amphetamines
Opiates
Phencyclidine

| Thoroughly plan for implementation |
| --- |
| <ul><li>Establish internal linkages (labor, EAP, management, union,...)</li><li>Assess pros/cons of potential units for organizational placement of drug testing (e.g., medical, personnel, EAP,...)</li><li>Determine organizational placement of drug testing (organizational unit)</li><li>Choose a certified laboratory</li><li>Disseminate notification to employees of implementation of drug testing program, provisions, right to confidentiality (have employees sign a policy acknowledgment form)</li><li>Determine sensitive positions (if any) for random testing, and send these individuals notification of such designation and that they may voluntarily admit to using illegal drugs (in which case they may be referred to the EAP). These individuals should be required to sign an acknowledgment</li><li>Secure a collection site</li><li>Train supervisors</li><li>Design supervisor report form to document specific events/behaviors observed which lead to testing</li><li>Begin employee education</li></ul> |

↓

## PLANNING FOR APPLICANT TESTING

| Determine goal for applicant testing |
| --- |
| To screen out applicants who are using illegal drugs, prior to being hired. This may have a positive effect on reducing instances of illegal drug use by employees |

↓

| Set up conditions for applicant testing |
| --- |
| <ul><li>Based on nature of workplace, determine extent of applicant testing</li><li>Indicate requirement of applicant testing in vacancy announcements of positions to be applicant-tested</li><li>Notify applicant when and where testing is to take place, and of confidentiality protections</li><li>Notify applicant that appointment to the position will be contingent upon a negative test result</li></ul> |

↓

## PLANNING FOR REASONABLE
## SUSPICION TESTING

↓

| Determine goal of reasonable suspicion testing |
| --- |
| To respond to documentable facts and circumstances leading to suspicion of illegal drug use, in order to protect the safety of co-workers and provide the suspected employee an opportunity for rehabilitation, in the case of a positive test result |

↓

| Set up conditions for reasonable suspicion testing |
| --- |
| • If an employee is suspected of illegal drug use, gather all information, facts, and circumstances leading to and supporting this suspicion<br><br>• Gain higher level concurrence, and prepare a written report detailing the basis for the suspicion<br><br>• Notify the employee |

| Reasonable suspicion = |
| --- |
| • Observable phenomena (actual use, possession, symptoms)<br><br>• Abnormal conduct or behavior<br><br>• Drug-related investigation, arrest, or conviction<br><br>• Employee drug test tampering<br><br>• Information from reliable sources |

↓

## PLANNING FOR ACCIDENT OR UNSAFE
## PRACTICE TESTING

↓

| Determine goal of accident and unsafe practice testing |
| --- |
| To provide a safe and secure work environment, by having the option to test any employee who is involved in an on-the-job accident, or who engages in unsafe on-duty job-related activities |

↓

| Specify conditions for accident and unsafe practice listing |
| --- |
| • Any employee involved in an on-the-job accident may be notified that testing will be initiated<br><br>• Specific criteria for accident or unsafe practice testing may include but are not limited to a death or personal injury requiring hospitalization, or damage to property in excess of a predesignated amount |

↓

# PLANNING FOR VOLUNTARY TESTING*

Determine goal of voluntary testing

To provide employees an opportunity to demonstrate their commitment to the goal of a drug-free workplace in their work setting and to set an example for other employees

Set up conditions for voluntary testing

- Any employee not in a position designated for random testing may volunteer to be included in the random testing pool, and be subject to the same conditions of those already designated

- Notify volunteers that they will remain in the random pool unless they officially withdraw from participation at least 48 hours prior to a scheduled test

* Only if random testing is part of drug-testing component.

# PLANNING FOR TREATMENT FOLLOW-UP TESTING

Determine the goal of treatment follow-up testing

To ensure that relapse has not occurred by employees during and after treatment

Set up conditions for treatment follow-up testing

- Notify all employees who undergo a counseling, treatment, or rehabilitation program for illegal drug use that they will be subject to unannounced testing following completion of such a program for a period of one year

- Determine the frequency of such testing (e.g., once a month)

# PLANNING FOR RANDOM TESTING

↓

| Determine goal of random testing |
|---|
| To eliminate specific risks involved with illegal drug use by employees in designated positions |

↓

| Set up conditions for random testing |
|---|
| • Ensure that persons in positions designated for testing receive ample notice of being subject to random selection<br><br>• Ensure that the means of random selection remains confidential<br><br>• Notify randomly selected individuals and their supervisors of scheduled testing with minimal advance notice (2 hours are recommended)<br><br>• The supervisor shall explain to the employee that she/he is under no suspicion and that her/his name was selected randomly<br><br>• A selected employee may obtain a deferral from testing if in a leave status or on official travel, but only with agreement from supervisors. (The employee may/should be rescheduled for testing within 60 days) |

↓

# SPECIMEN COLLECTION, LABORATORY ANALYSIS AND REPORTING, AND REVIEW OF RESULTS*

↓

| Purpose of proper procedures |
|---|
| To ensure careful and sensitive specimen collection procedure |
| • Make every effort to minimize the number of persons handling specimens through a preassigned chain of custody procedure.<br><br>• Allow individual privacy during collection unless there is reason to believe the specimen may be altered or substituted.<br><br>• Take all possible measures to ensure unadulterated specimens and correct identification of specimens.<br><br>• Maintain control and accountability of specimens by using a chain of custody form for each specimen from initial collection to final disposition.<br><br>• Have collection site personnel ship the specimens in a safe and secure manner to minimize the possibility of damage.<br><br>• If an employee refuses to be tested, remind him/her this could lead to disciplinary action, including termination |

* Details available in, "Mandatory Guidelines for Federal Drug Testing Programs".    ↓

→ | Ensure careful and sensitive laboratory analysis procedures

- Establish security measures to ensure that only authorized personnel have access to the specimens

- Require the laboratory to use both an initial and confirmatory test:

  - Initial immunoassay meeting FDA standards

  - Confirmatory for positive tests: GC/MS techniques

- Use only certified laboratories* which have qualified personnel and procedures

* Listing of certified laboratories is published monthly in the
*Federal Register.*

---

Ensure appropriate, thorough, and sensitive reporting and review of results

- The laboratory should report test results to the organization's Medical Review Officer (MRO) within 5 days, in a confidential, secure manner

- AII specimens negative on the initial test or negative in the confirmatory test must be reported as negative

- The MRO provides the final review of results, scrutinizing positive results for possible alternate medical explanations, conducting any necessary medical interviews with the employee, medical history reviews, or review of any other relevant factors

- If after the employee has a chance to discuss a positive result, the test result is verified as positive, the MRO then refers the case to the agency EAP as well as to the management official empowered to recommend or take administrative action

- The MRO may order a retest of a positive result should any question of accuracy or validity arise

## CONFIDENTIALITY OF TEST RESULTS
## AND RECORDS
↓

Ensure the maintenance of confidentiality when reporting results and concerning access to records maintained

- Results are confidential medical information which is not to be disclosed to any unauthorized individuals without the employee's consent

- Results must be kept on forms separate from employee records

- An employee who is tested may, upon written request, have access to records of the drug test (this does not include applicants)

- A record-keeping system should be devised maintaining the highest regard for employee privacy

↓

## QUALITY CONTROL PROCEDURES
↓

Ensure accuracy and reliability in your testing program through implementation and monitoring of quality control (QC) procedures

- Split portions of same specimen can be sent to the lab and should yield exactly the same results

- QC specimens that have been "spiked" with a drug and QC specimens that are certified to contain no drugs should be submitted to the lab along with employee specimen (a minimum of 3% of specimens submitted is recommended for good quality control)

- Close monitoring of QC specimen results will facilitate the detection of laboratory error

- The cost of maintaining QC programs should be recognized and planned for at the outset

↓

# DETERMINATION OF DISCIPLINARY ACTION AND DISPOSITION OF POSITIVE TEST RESULTS

If an employee receives a verified positive test result, the organization should refer the employee to the EAP, and if the employee occupies a sensitive position, the organization/company may choose to remove the employee from the position pending the employee's successful rehabilitation.

Applicant Testing:

- An applicant with a positive test result may be declined a final offer of employment

- The applicant may be allowed to reapply after a period of time (e.g., 6 months)

- The applicant should be informed of the reason for no job offer — a positive test result

All Other Types of Testing:

- An employee who tests positive should be referred to the EAP

- If the employee occupies a sensitive position, she/he may be immediately removed from the position; and, at the discretion of the employer, may be returned to duty in that position if it is determined that such return would not endanger the safety and security of others

- The employer may or may not take disciplinary action which could include removal/termination

- The severity of the action taken should depend on the circumstances of each case

## CHECKLIST FOR ESTABLISHING SAFEGUARDS

In developing a drug testing policy, balance the issues of employee right to privacy with principles of public safety.

Ensure a chain of custody process, through which the specimen route can be thoroughly documented. This will provide accountability in the event of legal scrutiny.

Ensure that all possible employee protections have been built into the program: ample and clear notice, confidentiality of records, professionalism throughout the process, and opportunity to contest the results.

Understand constitutional issues and recent court attitudes and decisions.

Have complete records of verification for positive test results.

Whenever possible, offer an employee who tests positive the chance for rehabilitation.

Purchase drug testing services only from a laboratory certified to ensure accuracy, validity, and reliability of test results.

# ■ APPENDIX D

## MODEL DRUG-FREE WORKPLACE
## PROGRAM DOCUMENTS

### WRITING A BASIC DRUG-FREE WORKPLACE POLICY

If you want only to establish a simple written drug-free workplace policy, a one-page model is provided on page 140 and a model cover letter is provided on page 141. You may copy them onto your company letterhead or, in consultation with your company's attorney, you may customize the policy to suit your company's special needs.

### WRITING A MORE COMPREHENSIVE POLICY

If you wish to write a more detailed policy, see the following pages for information and the appropriate paragraphs to add to your basic policy:

- Employee Assistance Program (EAP) (page 142; also see Chapter 8)
- Drug Testing (page 143; also see Chapter 7)
- Drug-Free Workplace Act Requirements (page 145; also see Chapter 9)

### EDUCATIONAL SERVICES TO HELP MAKE YOUR POLICY WORK

Education is one of the most vital components of any drug-free workplace program. Making employees aware of company policy as well as the consequences of policy violation is key to the overall strategy. A drug education program is also an opportunity to inform employees and their families of the adverse effects of illegal drug use on their health.

Education programs vary based on company size and resources. Your company newsletter may provide one means to bring information to employees' attention. Materials and speakers are frequently available at little or no cost from local community drug abuse prevention programs. The National Clearinghouse for Alcohol and Drug Abuse Information has a series of pamphlets and videotapes on "Drugs in the Workplace" designed for employee education available for loan. For additional resources contact the offices of the alcohol and drug abuse directors in your state (page 147).

If your company is subject to the requirements of the Drug-Free Workplace Act of 1988 (by nature of a grant/contract with the Federal Government) you should add the following statement to your drug policy:

> As a condition of employment, employees must abide by the terms of this policy and must notify (Company Name) in writing of any conviction of a violation of a criminal drug statute occurring in the workplace no later than 5 calendar days after such conviction.

## SAMPLE DRUG ABUSE POLICY STATEMENT

COMPANY
LETTERHEAD


DRUG ABUSE POLICY STATEMENT

(Company Name) is committed to providing a safe work environment and to fostering the well-being and health of its employees. That commitment is jeopardized when any (Company Name) employee illegally uses drugs on the job, comes to work under their influence, or possesses, distributes, or sells drugs in the workplace. Therefore, (Company Name) has established the following policy:

(1)   It is a violation of company policy for any employee to possess, sell, trade, or offer for sale illegal drugs or otherwise engage in the illegal use of drugs on the job.

(2)   It is a violation of company policy for anyone to report to work under the influence of illegal drugs.

(3)   It is a violation of the company policy for anyone to use prescription drugs illegally. (However, nothing in this policy precludes the appropriate use of legally prescribed medications).

(4)   Violations of this policy are subject to disciplinary action up to and including termination.

It is the responsibility of the company's supervisors to counsel employees whenever they see changes in performance or behavior that suggest an employee has a drug problem. Although it is not the supervisor's job to diagnose personal problems, the supervisor should encourage such employees to seek help and to advise them about available resources for getting help. Everyone shares responsibility for maintaining a safe work environment and co-workers should encourage anyone who may have a drug problem to seek help.

The goal of this policy is to balance our respect for individuals with the need to maintain a safe, productive, and drug-free environment. The intent of this policy is to offer a helping hand to those who need it, while sending a clear message that the illegal use of drugs is incompatible with employment at (Company Name).

## SAMPLE LETTER TO EMPLOYEES TO ACCOMPANY
## THE DRUG ABUSE POLICY STATEMENT

COMPANY
LETTERHEAD

LETTER TO ALL EMPLOYEES

The illegal use of drugs is a national problem that seriously affects every American. Drug abuse not only affects individual users and their families, but it also presents new dangers for the workplace.

The President of the United States has urged business and labor to take a leadership role in a nationwide effort to reduce the illegal use of drugs.

As you are aware, (Company Name) has always been committed to providing a safe work environment and fostering the well-being and health of our employees. Illegal drug use jeopardizes this commitment, and undermines the capability of (Company Name) to produce quality products and services.

To address this problem, (Company Name) has developed a policy regarding the illegal use of drugs that we believe best serves the interests of all employees. Our policy formally and clearly states that the illegal use of drugs will not be tolerated. This policy was designed with two basic objectives in mind:

(1)  Employees deserve a work environment that is free from the effects of drugs and the problems associated with their use, and
(2)  This company has a responsibility to maintain a healthy and safe workplace.

I believe it is important that we all work together to make (Company Name) a drug-free workplace and a safe, rewarding place to work.

Sincerely,


President
(Company Name)

## BASIC INFORMATION ON EMPLOYEE ASSISTANCE PROGRAMS (EAP):

Emotional problems, marital and family discord, financial or legal difficulties, alcoholism and drug abuse are usually considered "personal" problems until they affect the individual's job performance. Then they become your company's problem, and you need an effective solution.

For many employers, that solution is an employee assistance program (EAP), a resource to which employees and their families can turn for confidential, professional assistance.

An EAP establishes a way for troubled employees to seek help directly or for supervisors to refer those employees whose problems are affecting job performance. The employee assistance professional conducts a thorough, confidential assessment and then provides short-term therapy or refers the employee to an appropriate therapist, treatment program, or community agency.

As a component of your drug-free workplace program, the EAP should provide training for supervisors and managers, as well as education to make your employees aware of the assistance program and how it works.

For further information call:

### National Resources

- National Institute on Drug Abuse (NIDA), Workplace Helpline 1-800-843-4971 (Toll Free)
- Employee Assistance Professionals Association (EAPA): 4601 North Fairfax Drive, Suite 1001 Arlington, VA 22203. Telephone: 703-522-6272. (EAPA provides information on how to select EAPs, the value they can provide, the theory behind them, and how they operate.)

### Local Resources

(Use this space to list local resources as you identify them.)

If you are including an EAP in your company's program, add the following paragraph to your policy:

> The company offers an Employee Assistance Program (EAP) benefit for employees and their dependents. The EAP provides confidential assessment, referral, and short-term counseling for employees who need or request it. If an EAP referral to a treatment provider outside the EAP is necessary, costs may be covered by the employee's medical insurance, but the costs of such outside services are the employee's responsibility.

## BASIC INFORMATION ON DRUG TESTING AND RESOURCES

According to a 1990 study by the Business Round Table, the vast majority of large U.S. employers use drug testing in certain circumstances.

Drug testing is primarily intended to protect employees' health and safety through the early identification and treatment of alcohol and other drug abuse problems.

Policies regarding pre-employment and employee testing are often determined by the risks associated with safety, security, and health. Many employers test when there is "reasonable suspicion" or "probable cause" to believe an individual is using drugs. Others conduct routinely scheduled testing. Still others feel that universal testing is warranted.

Many unions have negotiated testing agreements and have taken a strong position condemning drug use by union members.

A few states have passed laws restricting workplace drug testing, most notably Connecticut, Iowa, Montana, Rhode Island, Maine, and Vermont. You should check with your company's attorney or State Attorney General (see list on page 146 under State Laws) regarding the specific laws that apply to your area.

If you choose to test applicants/employees, it is strongly recommended that you use a laboratory certified for drug testing. The National Institute on Drug Abuse (NIDA) certifies laboratories for federally mandated drug testing. NIDA-certified laboratories are listed in the *Federal Register* on or about the first of every month. Some states also certify laboratories for drug testing. It is further recommended that strict procedures should be followed for supervising the chain of custody of samples and the medical review of test results (see Chapter 7).

For further information call:

## National Resources

- National Institute on Drug Abuse (NIDA)
  Toll-Free 1-800-843-4971
- National Clearinghouse for Alcohol and Drug Information
  Toll-Free 1-800-729-6686

## State Resources

- Telephone numbers for State Attorneys General (page 146 under State Laws)
- Telephone numbers for National Association of State Alcohol and Drug Abuse Directors (page 147)

If you are adding drug testing to your drug-free workplace policy, add paragraph #1 (pre-employment) and/or paragraph #2 (employment).

1. Pre-employment drug testing paragraph:

All job applicants at this company will undergo testing for the presence of illegal drugs as a condition of employment. Any applicant with a confirmed positive test result will be denied employment. This company will not discriminate against applicants for employment because of a past history of drug abuse. Therefore, individuals who have failed a pre-employment test may initiate another inquiry with the company after a period of no less than 6 months, but must present themselves drug-free.

2. Employee testing paragraph:

This company has adopted testing practices to identify employees who use illegal drugs either on or off the job. It shall be a condition of employment for all employees to submit to drug testing under the following circumstances:

- When there is reasonable suspicion to believe that an employee is using illegal drugs.
- When employees are involved in on-the-job accidents where personal injury or damage to company property occurs.
- As part of a follow-up program to treatment for drug abuse.

## BASIC INFORMATION ON THE LAW

### Drug-Free Workplace Policy Law

Federal, state, and local law pertaining to drug-free workplace policies and procedure is complex and subject to frequent change by legislation and court decisions. It is essential to contact your company's attorney before implementing any policy, practice, or change. This book and these materials do not constitute legal advice.

Through the local chapter of the American Bar Association, some attorneys in your community may be available to answer without charge specific questions regarding drug-free workplace policies.

### Federal Regulations

The Drug-Free Workplace Act of 1988 requires all federal grant recipients and federal contractors (where contracts exceed $25,000) to certify that they will provide a drug-free workplace. The final rules describing the requirements for such grantees/contractors were published in the *Federal Register* on May 25, 1990.

Generally the law requires covered employers to:

- Develop and publish a written policy and ensure that employees read and consent to the policy as a condition of employment
- Initiate an awareness program to educate employees about
  - The dangers of drug abuse
  - The company's drug-free workplace policy
  - Any available drug counseling, rehabilitation, and employee assistance programs

- The penalties that may be imposed upon employees for drug abuse violations
- Require that all employees notify the employer or contractor of any conviction for a drug offense in the workplace
- Make an ongoing effort to maintain a drug-free workplace.

If your company is subject to the requirements of the Drug-Free Workplace Act of 1988 (by nature of a grant/contract with the federal government) you should add the following statement to your drug policy:

> As a condition of employment, employees must abide by the terms of this policy and must notify (Company Name) in writing of any conviction of a violation of a criminal drug statute occurring in the workplace no later than 5 calendar days after such conviction.

## Department Of Transportation (Dot) Regulations

The U.S. Department of Transportation (DOT) rule on drug testing regulations became effective in December 1988. The regulations cover several occupations under DOT jurisdiction, including natural gas and pipeline workers, motor carrier workers, aviation workers, and railroad workers. Employers with transportation positions covered by DOT must test job applicants. Employees are to be tested during routine physicals, on a random basis, upon reasonable cause, and after accidents.

The DOT has established an Anti-Drug Information Center (ADIC); this computer-based system can respond to voice telephone calls, facsimile, or modem. The system will provide callers with model drug rules and detailed information, interpretation, and advice on DOT regulations. Contact 1-800-CAL-DRUG.

## The Americans With Disabilities Act Of 1990

This Act, effective July 1992, prohibits discrimination against "qualified people with disabilities" and limits an employer's ability to inquire into an employee's or job applicant's medical history. It does, however, permit drug testing and does not bar employers from prohibiting alcohol abuse or illegal drug use in the workplace.

Although the Act does not protect certain illegal substance abusers and alcoholics who cannot safely perform their jobs, it does protect those who have been rehabilitated or who are participating in supervised rehabilitation programs and not currently using drugs.

The regulations appeared in the July 26, 1991 *Federal Register.*

## State Laws

State laws vary. Some states have passed legislation restricting drug testing. You need to be aware of what is required in your state. The Attorney General's office in your state capitol can give you the regulations that pertain to your state. Refer to the following list if you need information.

# OFFICES OF STATE ATTORNEYS GENERAL

| | |
|---|---|
| Alabama | 205-242-7300 |
| Alaska | 907-465-3600 |
| Arizona | 602-542-4266 |
| Arkansas | 501-682-2007 |
| California | 916-445-9555 |
| Colorado | 303-620-4500 |
| Connecticut | 203-566-2026 |
| Delaware | 302-577-3838 |
| District of Columbia | 202-727-6248 |
| Florida | 904-487-1963 |
| Georgia | 404-656-4585 |
| Hawaii | 808-586-1282 |
| Idaho | 208-334-2400 |
| Illinois | 217-782-1090 |
| Indiana | 317-232-6201 |
| Iowa | 515-281-5164 |
| Kansas | 913-296-2215 |
| Kentucky | 502-564-7600 |
| Louisiana | 504-342-7013 |
| Maine | 207-289-3661 |
| Maryland | 301-576-6300 |
| Massachusetts | 617-727-2200 |
| Michigan | 517-373-1110 |
| Minnesota | 612-296-6196 |
| Mississippi | 601-359-3680 |
| Missouri | 314-751-3321 |
| Montana | 406-444-2026 |
| Nebraska | 402-471-2682 |
| Nevada | 702-687-4170 |
| New Hampshire | 603-271-3658 |
| New Jersey | 609-292-4925 |
| New Mexico | 505-827-6000 |
| New York | 212-341-2519 |
| North Carolina | 919-733-3377 |
| North Dakota | 701-224-2210 |
| Ohio | 614-466-3376 |
| Oklahoma | 405-521-3921 |
| Oregon | 503-378-6002 |
| Pennsylvania | 717-787-3391 |
| Puerto Rico | 809-721-2900 |
| Rhode Island | 401-274-4400 |
| South Carolina | 803-734-3970 |
| South Dakota | 605-773-3215 |
| Tennessee | 615-741-3491 |
| Texas | 512-463-2100 |

| | |
|---|---|
| Utah | 801-538-1015 |
| Vermont | 802-828-3171 |
| Virgin Islands | 809-774-5666 |
| Virginia | 804-786-2071 |
| Washington | 206-753-6200 |
| West Virginia | 304-348-2021 |
| Wisconsin | 608-266-1221 |
| Wyoming | 307-777-7841 |

## OFFICES OF STATE ALCOHOL AND DRUG ABUSE DIRECTORS

| | |
|---|---|
| Alabama | 205-270-4650 |
| Alaska | 907-586-6201 |
| Arizona | 602-255-1025 |
| Arkansas | 501-682-6650 |
| California | 916-445-1943 |
| Colorado | 303-331-6530 |
| Connecticut | 203-566-4145 |
| Delaware | 302-421-6102 |
| District of Columbia | 202-673-7481 |
| Florida | 904-488-0900 |
| Georgia | 404-894-6352 |
| Hawaii | 808-586-3962 |
| Idaho | 208-334-5935 |
| Illinois | 217-785-9067 |
| Indiana | 317-232-7816 |
| Iowa | 515-281-3641 |
| Kansas | 913-296-3925 |
| Kentucky | 502-564-2880 |
| Louisiana | 504-342-9354 |
| Maine | 207-289-2595 |
| Maryland | 301-225-6925 |
| Massachusetts | 617-727-7985 |
| Michigan | 517-335-8808 |
| Minnesota | 612-296-4610 |
| Mississippi | 601-359-1288 |
| Missouri | 314-751-4942 |
| Montana | 406-444-2827 |
| Nebraska | 402-471-2851 |
| Nevada | 702-687-4790 |
| New Hampshire | 603-271-6104 |
| New Jersey | 609-292-5760 |
| New Mexico | 505-827-2601 |
| New York (Alcohol) | 518-474-5417 |
| New York (Drugs) | 518-457-7629 |
| North Carolina | 919-733-4670 |
| North Dakota | 701-224-2769 |

| | |
|---|---|
| Ohio | 614-466-3445 |
| Oklahoma | 405-271-8777 |
| Oregon | 503-378-2163 |
| Pennsylvania | 717-787-9857 |
| Puerto Rico | 809-764-3795 |
| Rhode Island | 401-464-2091 |
| South Carolina | 803-734-9520 |
| South Dakota | 605-773-3123 |
| Tennessee | 615-741-1921 |
| Texas | 512-867-8802 |
| Utah | 801-538-3939 |
| Vermont | 802-241-2170 |
| Virgin Islands | 809-773-1992 |
| Virginia | 804-786-3906 |
| Washington | 206-753-5866 |
| West Virginia | 304-348-2276 |
| Wisconsin | 608-266-3442 |
| Wyoming | 307-777-7115 |

# ■ APPENDIX E

## THE EMPLOYEE ASSISTANCE PROGRAM

### Setting up your program

The policy you adopt and the programs you choose to implement that policy will be determined by many factors including the size of your business, your product or service line, your relationships with your employees, whether your employees are represented by a union, your medical insurance coverage and other fringe benefits, the difficulty and costs of replacing trained employees, the laws affecting your business, your personal attitudes toward substance abuse, and your role in addressing the needs of society.

Your policy should be written, and it must be made known. Large employers usually publish a detailed handbook for employees, setting forth policy, procedures, and benefits of the company. You may wish to publicize your substance abuse policy in the same way as other policies are made known in your organization (letters, memoranda, notices, or employee handbooks or manuals).

A policy that clearly delineates unacceptable behavior and its consequences can result in benefits for all involved. For the employer, such a policy saves management time, reduces legal liability and other risks, and minimizes surprises. Employees cannot say that they did not know a particular behavior was prohibited or what the consequences of such behavior would be. Employees, for their part, gain freedom from a clear policy because they know precisely what behavior is proscribed and what the consequences of breaking the rules are. The chance of discrimination or preferential treatment among the workers is minimized.

In order to obtain these results, your policy must set forth precisely what you, the employer, consider substance abuse to be and what actions you will take in response to substance abuse (all actions need not be punitive), and what provisions you make to help employees keep or retain good health as well.

### A Successful Policy Should Be

- Consistent with your organization's values and other policies
- Backed up with workable procedures
- Designed to create only those expectations that can and will be fulfilled
- Respectful of employees' rights, while protecting the organization
- Consistent with current drug testing, assistance, and rehabilitation programs
- Executed equally for all employees

Policies adopted by other businesses range from getting rid of the employee through immediate dismissal ("zero tolerance") to eliminating the abuse by providing assistance to the employee in overcoming his or her addiction ("safe harbor").

### Zero Tolerance

A zero tolerance policy states that any employee found to possess, distribute, or use illegal drugs, whether on or off the job, will be terminated. Alcohol and prescription drug abuse also might be included, particularly when the operation of dangerous equipment or motor vehicles is involved. Identification may occur through testing, searches, direct observation of use or possession, or conviction. There is no chance for being rehired, even with rehabilitation.

There are benefits to such a policy. Everyone is on notice that employees using drugs must stop and that no drug users need apply for employment. If pre-employment and for-cause testing are included in the policy, the statement is even stronger. The employer does not have to incur the cost of rehabilitation, treatment, or the risk of retaining or reinstating an unreliable employee.

There are situations and circumstances where immediate dismissal is necessary and appropriate, but successful implementation of this policy requires careful consideration of the legal ramifications, the possibility of lawsuits for wrongful discharge, the difficulty of proving substance abuse, setting up secure testing programs, deciding what to do about employees who voluntarily seek help, estimating the costs of possibly losing good employees, and other factors. The strict enforcement of this policy does not allow an opportunity for employees to seek help voluntarily, and may cause employees to hide their problems. A middle ground implemented by many employers permits assistance through confidential self-referral to get help and stop using drugs before getting caught. Immediate discharge may be only one option among a variety of alternatives.

## ■ SAFE HARBOR

Under this philosophy, employees — even those who test positive for drugs — are dealt with sympathetically, with their substance abuse being seen as an illness. They may return to work following rehabilitation; if they experience a relapse, they may return to the job and try again. Such a policy may save employees' lives and return good employees to productivity. It is clear that the employer will be seen as supporting the efforts of employees who conscientiously want to be rehabilitated, and the company will be seen as participating in efforts to combat substance abuse.

However, there are risks to the safe harbor policy. The employer may end up, in effect, allowing an employee to perpetuate the problem. If the employee fails to become free from drug abuse, the employee has not really been "helped". It's also costly to invest and fail; jobs may be held open for employees who never return to full productivity. Without appropriate training and defined procedures, managers may be seen to be allowing substance abuse to continue.

## Policy Checklist

The following list suggests the scope of questions that need to be considered when designing substance abuse policies and programs:

- What government regulations concerning the workplace must be addressed?
- What provisions of general or specific law must be addressed?
- What is the nature of the organization's and employees' work? Do employees' duties involve classified information? Could their duties be dangerous to personal or public safety if performed by an impaired person?
- Is there a union involved?
- Shall the focus on substance abuse be on the extremes of zero tolerance or safe harbor, or some middle course?
- Is there, or will there be, drug testing? If so, what kinds — pre-employment, for-cause, or random?
- What security and safety measures related to drugs, including searches, if any, will be implemented?
- What resources will the company provide in terms of education, training, access to EAP, and benefits coverage? The program should indicate what services are available and who pays for them, and the extent of confidentiality.
- What statement about the company does the employer wish to convey to its current and prospective employees?
- What price is the company paying for not having a policy, program, or procedures?

Increasingly, large and small employers are establishing EAPs as management tools to help their employees deal with substance abuse and other personal and professional problems.

## What the EAP is

The term employee assistance program (EAP) refers to a formal structured service provided and paid for by the employer.

An EAP is a worksite-promoted program designed to assist in the identification and resolution of productivity problems associated with employees impaired by personal concerns including, but not limited to: health, marital, family, financial, alcohol, drug, legal, emotional, stress, or other personal concerns which may adversely affect job performance.

EAPs include counseling services designed to help employers identify and cope with such problems. These services are both a management resource providing advice, guidance, and training on a variety of employee relations issues, and an employee resource providing counseling on individual personal problems.

## The Basic Components of an EAP Are

- A policy statement that defines the employer's position as to substance abuse, the kinds of problems covered, how employees get help, and a commitment to confidentiality.
- Procedures for case handling that define program parameters, confidentiality, and how client contact is made.
- Accountability to ensure the program:

  - Is consistent with labor and management policies
  - Defines roles
  - Supports sound employer practices
  - Provides for evaluation

- Identification of treatment resources, with an emphasis on appropriate referral for the client and effective and cost-efficient service.
- Communication to ensure the EAP is visible and understood at all levels
- 24-hour, 7-day access by telephone or face-to-face to an EAP counselor
- Training of management and workers to deal with employee problems
- Confidentiality in record keeping and counseling
- Appropriate insurance coverage to provide adequate coverage for substance abuse treatment in a cost-efficient manner.

## Services Provided by an EAP

Services provided may be either broad or limited. Comprehensive EAPs address nearly every personal problem that creates a negative impact at work including productivity, security, or safety. Good broad-based programs effectively address costly personnel problems, including substance abuse.

EAPs that are limited to providing only counseling on substance abuse problems are unusual. Although substance abuse is a major problem for many employers, EAPs report that as many as four out of five problems brought to counselors involve family and work stresses. Because most employees do not, and frequently cannot, admit they are abusing substances, this may not come to light until they seek help for another related problem.

The services usually provided by an EAP include, but are not limited to the following:

- Consultation for management and policy development
- Training for supervisors and managers

- Access to a professional counselor at any time
- Assessment of problems
- Referral to external treatment or other community services, if necessary
- Counseling to encourage problem resolution
- Counseling for employee family members
- Case follow-up
- Employee education
- Employee information on EAP availability and services
- Information on community resources for substance abuse and other problems
- Seminars on health-related topics
- On-site visits by an EAP counselor
- Optional service locations, times, and methods
- Assurance of total confidentiality except when there is a legal "duty to inform"
- Reports on EAP utilization (without breaching confidentiality)

## Related Services

Additional EAP components may include:

- Coverage of employees' family members, including confidential self-referral by a covered family member — an important resource to reveal employee problems as well
- Brief counseling (from three to eight sessions in addition to sessions required for assessment)
- On-site visits by the EAP counselor to deal with situations before they become problematic, to meet with supervisors, to encourage use of the EAP's services, and to build trust
- Optional service locations, times, and methods. Options can include counseling at the worksite or in one or more counseling offices off the worksite, by telephone or face-to-face, or at various times to accommodate shift workers

The following services may be included in a comprehensive EAP or purchased from separate vendors:

## Managed Care Components

To hold down costs of providing mental health and substance abuse benefits, some employers contract with a managed care company to provide pre-certification, utilization review, and case management. The EAP may be integrated into this process as a "gatekeeper" to perform some of the pre-certification. It may also provide case management functions that go beyond traditional "EAP case management".

This approach is intended to prevent inappropriate "treatment". At its best, managed care structures good quality, cost-attentive medical care with specially trained practitioners as gatekeepers and case managers. Thus, it may benefit employer and employees. Many EAP providers offer managed care as an additional service.

However, there are disadvantages and risks. There are, for example, additional up-front costs to the employer. Managed care also may result in reduced quality or quantity of care, insufficient inpatient time for detoxification or treatment, and insufficient rehabilitation resulting in a higher relapse rate. Comprehensive EAP contracts with external providers should include "EAP case management" which performs some of the "managed care" functions such as pre-certification of providers. It is important to review both plans to ensure that the employer is not paying twice for the same or similar service. The EAP should be regarded as the employee's advocate in dealing with treatment providers. In some cases, the managed care involvement may create conflicts of interest.

## Wellness and Health Promotion

Broad EAPs are concerned with the overall health of the employees. They offer seminars on such health-related topics as AIDS, weight control, reduction or elimination of smoking, back injuries, hearing injuries, and management of stress, and encourage use of EAPs.

## Counseling

While some large employers may have EAP counselors on their staff, most obtain these services from outside providers. These providers range in size from a single social worker providing all the services to large corporations.

### What Is Counseling?

Specially trained counselors on the provider's staff "assess" problems, provide family outreach, and are familiar with community resources. The focus is on coaching, guiding, training, and motivating clients to see options and change certain behaviors. The counselor may provide brief counseling for problems that can be resolved in a few sessions, or may refer an individual to external specialists for diagnosis, treatment, or legal, career, financial, or other help as required.

The assessment process is critical. More than half of employees who use EAP services present themselves as having "family problems". For many, the family problems are related to substance abuse by themselves or their mates or children. Substance abuse problems may be identified in a third or more of these cases if the counselor is experienced.

It should be noted that there is a gray area where counseling leaves off and therapy begins. When to counsel and when to refer for treatment depends on the nature and severity of the problem and the projected length of time required for resolution. Substance abuse cases should be referred to external specialists, with the EAP counselor following up.

The counselor should be expected to continue to follow the case to ensure that satisfactory resolution is achieved where substance abuse is involved. Ongoing follow-up is essential. Counselors also are responsible for consulting with employees and supervisors about helping affected employees continue to work or return to work following treatment.

### Where Is Counseling Provided?

EAP services may be provided at the worksite or off-site at the provider's offices. Some EAPs offer 24-hour telephone hotline counseling for either information and referral, or brief counseling and periodic contact for support and follow-up. There are advantages and disadvantages to each of these alternatives. However, small employers rely on off-site telephone or face-to-face assistance, through external service providers.

The use of telephone counseling services rather than exclusively requiring face-to-face meetings with the counselor is controversial. At a minimum, this type of counseling requires professionally trained personnel experienced in telephone counseling. Cases requiring treatment for substance abuse should be seen face-to-face, prior to referral except for emergency situations.

### How Is Treatment Provided?

EAP counselors do not provide treatment. They recommend referral options when treatment is necessary. They may refer to free community resources such as Alcoholics Anonymous (AA), Narcotics Anonymous, Cocaine Anonymous, Co-Dependents Anonymous, Adult Children of Alcoholics, and other self-help programs, or to mental health, legal, financial, career, or other specialists depending on the identified problem; or, they may refer for professional treatment at either a nonprofit or private clinic, hospital, or other treatment facility. A combination of multiple approaches is common.

### What are the costs of an EAP?

Smaller employers may contract directly with any of the numerous EAP providers that offer their services to more than one client, may choose to enter into a cooperative arrangement with other small businesses in the same area or same product line, or may help his or her employees establish an employee-operated peer-based program.

Costs of an EAP are based on such variables as the extent of services provided, number of employees, number of sites, type of industry, specific regulatory requirements, the employer's policy regarding drug testing, and whether dependents are covered. Most external EAP providers serve more than one client company, so overhead costs are distributed among those clients. The rule of thumb calls for one counselor for each 3500 to 5000 employees.

Fees for EAP services vary considerably among providers. Most commercial EAPs contract for a designated amount per employee per year or a flat annual fee. All have price differentials with significantly higher rates for small employers particularly, for those with fewer than 100 employees. Some EAP providers have additional fees for materials, administration, and start-up. Minimum base fees range up to $7,500.

## Fixed-Fee Contracts

Fixed-fee contracts are the most common approach. Fees are calculated on a per-employee-per-year basis, and the provider delivers the agreed-upon services regardless of the extent of utilization. The amount of the fee is determined by the number of counseling sessions stipulated in the contract, face-to-face counseling requirements, the need for counselors to be available to cover split shifts and nighttime staff, cost of materials, administrative costs, and profit margin.

In 1993, average costs range from $25 to $45 per employee per year (sometimes quoted at $2.00 to $3.50 per employee per month) for businesses with more than 250 employees. For organizations with fewer than 100, the rates may range up to $100 per employee per year or a flat fee of $2000 to $5000 per year, regardless of number of employees under the ceiling of 100.

In 1993, a typical broad-based program with up to three sessions for assessment and brief counseling would cost from $22,000 to $30,000 per year ($22 to $30 per employee) for an organization with 1000 employees, while a company of fifty might be expected to pay anywhere from a minimum of $2500 up to $7500 ($50 to $75 per employee). For this per-employee fee, the employer would receive all of the basic services of a broad-based program, with coverage of family members.

Due to competition, some EAP firms are cutting prices by eliminating on-site training, management consulting, and visitations, and other labor-intensive services or those that tend to increase utilization. Under competitive pressures, a fixed-fee contract for minimal EAP services may be as low as $5000 or $10 per employee, for a firm of 500 employees. For firms of 50 employees, quotes may be as low as $500.

## Consortia

A consortium is an organization which small employers can join to obtain the services of an EAP provider. A consortium may be arranged through the initiative of one or more individuals or through trade or civic organizations. A provider may directly approach several businesses and suggest forming a consortium using that provider's services. Costs per employee are frequently reduced by guaranteeing the provider a sufficiently large employee-client base to allow the same economies of scale achieved by a single large company. In this model, the consortium acts as a service to employers by contracting directly with providers for employee services at specific rates and monitors the services to assure employers are obtaining quality services as needed. In the metropolitan Washington area, the Corporation Against Drug Abuse (CADA) has established a Small Employer Consortium; information on this organization is available at (202) 338-0654.

In general, the services provided through a consortium are not individually tailored to the unique circumstances of each small business member. Training may be provided to groups of supervisors or employees from several companies at the same time, individual management counseling may not be as readily available, on-site visitations by the counselor may be limited, and the same materials may be used to publicize EAP services.

Costs to members of a consortium can be as low as $36 to $50 per employee per year. Thus, a company of 50 employees can have access to EAP services as low as $1800 per year.

**Fee-for-Service Programs**

In a fee-for-service program, the employer contracts directly with qualified specialists for the specific type of service required, and pays only for the service actually provided. This allows the employer to determine which services (counseling, referral, training, etc.) appear to be needed and know the cost of that specific service. It requires the employer to make individual decisions about each referral for counseling or requests for other services and may make confidentiality more difficult to ensure.

For the employer, program costs are very low if utilization is low. In turn, the provider, paid for each contact hour, has no risk. Administrative start-up costs and reporting fees are low; there can be provision for a flat hourly or per-project fee for training.

The actual cost to the employer depends upon the number of employees who use EAP services, the hourly rates for the services used, and the number of related services (e.g., training, management consultation, on-site visitation, etc.) obtained by the employer. The annual cost for a company with 50 employees may be about $1000, if few employees need counseling and training and related services are shared.

**Physician-Based Services**

Primary care physicians (family physicians, general practitioners, group practices) may offer EAP services as an adjunct to their medical practice. These physicians offer employee assistance services (problem assessment, counseling, referral to treatment, case management and follow-up, supervisor and employee training, and reports) for any health-related problem, including substance abuse. In addition, physician-based services may provide medical examinations, arrangements for drug tests and urine collection, and occupational medicine services.

The employer receives a list of specially trained physicians, which they make available for supervisor referrals and for employees who self-refer. Physician fees, typically $85 to $150, are paid on a fee-for-service basis, by the employer in supervisor-referred cases or by the employee in self-referred cases. Any medical insurance payments are assigned to the party that paid the fee.

This approach is appropriate for very small businesses because there are no up-front costs. Part of the costs that are incurred may be reimbursed through existing medical benefits. Further, the employer may gain agreement from the physicians that they will specify referral alternatives that are consistent with insurance coverage. The infrastructure exists at no extra charge if the employer provides any medical benefits.

The services may be less threatening and better utilized for early intervention as the employee or family member is seen as having a medical problem rather than a substance abuse problem. For supervisor referrals, this physician-based plan reduces the risk of implications that a troubled employee has substance abuse problems. Another benefit of this plan is that laboratory tests, including urine screening, are available from the same source as part of an "assessment" or following accidents. The laboratory tests may be part of a more comprehensive physical examination or specifically related to drug testing protocols.

### Peer-Based Programs

Some employee assistance services incorporate co-worker (peer) involvement. Based on "management bypass" agreements, the peer-based (or co-worker-based or, in the case of unions, member-based or member assistance program) system allows the employees to become agents of change. They turn the culture of an organization from one of permissiveness, denial, and enabling behaviors of substance abuse to one of zero tolerance with compassion and support.

The peer-based program includes peer prevention through education and training, peer intervention (identification and confrontation), peer referral, and peer support. This option requires a great deal of formal education and training. For small employers in which this type of EAP is not provided by a union, it is advisable to obtain the required consultation and training services through a consortium or cooperative group.

The great benefit of this option is that co-workers generally identify problems before supervisor, family members, and even the impaired employees become fully aware of the problem's presence or severity. Co-workers can be part of the problem or part of the solution.

### "No-Cost" Services

Some hospitals, treatment centers, and "preferred provider organizations" (providers of psychiatric and substance abuse treatment) are offering free employee assistance services to employers. Although some of these services may resemble contracted, independent EAPs, they are different in that their major reason for providing free services is to develop good will and to promote their own treatment facilities. There are benefits to such services. Specialists are available because they are affiliated with a hospital or treatment center. There may be free access to assessment and crisis sessions, and free training for supervisors.

However, there also are risks. Free service is a loss leader to develop business for the hospital or treatment center. The institutions may not be unbiased; they may be less inclined to refer individuals to free-standing, community-based services rather than providing services for which they will be paid. The purpose of an EAP is to serve the best interests of the employer and the employees. This means guiding employees who need help into the most appropriate and least costly treatment and facility. The institution offering the EAP with no front-end costs may not be the best treatment resource for the problem or patient.

Also, because a treatment facility benefits directly from referrals, there may be excessive and perhaps not absolutely necessary referrals for treatment that will be paid for by the employee, with or without reimbursement from insurance. This can escalate insurance rates for the employer and may discourage the referred employee from entering into or completing treatment due to prohibitive costs. At the same time, treatment provided by such institutions may be as good as or better than that provided by other facilities in the community, and the costs may be similar or lower.

# GLOSSARY

## DRUG TERMINOLOGY AND SLANG TERMS

In order to better understand the drug culture, it is sometimes important to note the different terms used for the drugs or paraphernalia. Below are some of the slang terms users speak, although the terms vary from one region to another across the country:

### A

| | |
|---|---|
| **A-Bomb** | A mixture of heroin and marijuana. |
| **Acapulco Gold** | Marijuana. |
| **Ace** | Marijuana cigarette. |
| **Acid** | Lysergic acid (LSD). |
| **Acid head** | LSD user. |
| **Acid rock** | A type of music considered to be compatible with LSD-induced states of consciousness. |
| **Acid test** | Party at which LSD has been added to the punch. |
| **Action** | Selling of drugs. |
| **A-head** | Amphetamine user. |
| **A.K.S.** | Look-alike drugs containing caffeine and phenyl-propanolamine sold on the street as speed. |
| **Alice B. Toklas cookies** | Cookies made from marijuana. |
| **Amobarbital** | Amytal. |
| **AMT** | Amphetamines. |
| **Amyl nitrate** | A vasodilator, illegally snorted to give excitement; sold over the counter as "rush". |
| **Amys** | Vials of amyl nitrate. |
| **Angel dust** | Phencyclidine hydrochloride, PCP. |
| **Angel off** | Police term; to arrest the customers of a drug dealer. |
| **Anywhere** | Possessing drugs. |
| **Army disease** | During the Civil War the injection of morphine became possible through the invention of the hypodermic needle. Name used to describe the soldiers addicted to morphine. |
| **Artillery** | Equipment for injecting drugs. |

| | |
|---|---|
| **Ataraxic drug** | Any drug that acts like a tranquilizer. |
| **Atropine** | An alkaloid found in deadly nightshade. Used as an antispasmodic medication. |
| **Attitude** | Sudden hostile feelings. |
| **Ayahuasca** | Peruvian Indian word for a hallucinogenic beverage. Also called Caapi, Natema, and Yage. |

## B

| | |
|---|---|
| **Baby** | Marijuana. |
| **Back up** | Injecting heroin into the vein, then withdrawing part of the blood back into the syringe. |
| **Bad** | Refers to very good, high in potency. |
| **Bad seed** | Peyote. |
| **Bad trip** | A panic or crisis reaction following taking LSD or hallucinogens. |
| **Bag** | Small packet of drugs. |
| **Bagman** | Supplier of drugs, or a holder of money for an illegal transaction. |
| **Bale** | A pound of marijuana. |
| **Balloons** | Heroin or other drugs sold in rubber balloons so that, if the transporter faces pending arrest, he/she can swallow the balloon and later retrieve the balloon and use the contents. |
| **Bam** | An amphetamine pill. |
| **Bambu** | A kind of cigarette paper, normally used to roll marijuana cigarettes. |
| **Bang** | Injection of narcotics. |
| **Bank Bandit Pills** | Barbiturates or other sedative type pills. |
| **Banker** | One who finances the purchase of drugs. |
| **Bar** | A block of marijuana, usually bound together with sugar or honey water. |
| **Barbs** | Barbiturates. |
| **B-bomb** | Benzedrine inhaler. |
| **B.C.** | Birth control pill. |
| **Beans** | Amphetamines; also used for mescaline. |
| **Beast** | LSD. |
| **Beat** | To rob. |

| | |
|---|---|
| **Bee** | A measure of sale, usually enough marijuana to fill a matchbox. |
| **Behind acid** | Using LSD. |
| **Beinsa** | A plant, chewed or drunk as a tea, thought to have a narcotic effect. |
| **Belladonna** | A drug containing alkaloids of atropine, scopolamine, and hyoscyamine. |
| **Belly habit** | A gnawing in the stomach caused by the use of opiates. |
| **Belt** | The high following the ingestion of a drug. |
| **Belted** | High. |
| **Bennies** | Benzedrine tablets, also referred to as "uppers". |
| **Bent** | High or intoxicated from a hallucinogen or other drug. |
| **Benz** | Benzedrine pills. |
| **Bernice** | Cocaine. |
| **Bernies** | Cocaine. |
| **Bhang** | Hashish. |
| **Big bags** | Five $10.00 bags of heroin. |
| **Big C** | Cocaine. |
| **Big Chief, The** | Mescaline or peyote. |
| **Big D** | LSD. |
| **Big H** | Heroin. |
| **Big Man, The** | The number one dealer in narcotics; generally, the wholesaler who supplies the drugs to the pusher. |
| **Big Supplier** | The big man. |
| **Bindle** | Small amount of drugs packaged in folded paper or a glassine envelope. |
| **Bing** | Solitary confinement in jail, or an injection of narcotics. |
| **Bit** | Serving time in jail |
| **Biz** | Equipment for injecting drugs |
| **Black Beauty** | Black capsule, look-alike drug containing caffeine and phenylpropanolamine sold on the street as speed. |
| **Black Birds** | Amphetamines. |

| | |
|---|---|
| **Black Gunion** | An extra thick, dark gummy type of marijuana. |
| **Black Mollies** | Amphetamines. |
| **Black Pills** | Pellets of opium used for smoking. |
| **Black Russian** | Hashish. |
| **Black Stuff** | Opium. |
| **Black tar** | Heroin. |
| **Blank** | Container of non-drugs, such as talcum powder or sugar, sold to an addict. |
| **Blanco** | Spanish for white heroin. |
| **Blast** | To smoke marijuana. |
| **Blast party** | A party to smoke marijuana. |
| **Blaze** | LSD or ACID. |
| **Blind munchies** | Strong appetite for sweets, part of the marijuana high. |
| **Block** | Barbiturates, or a cube of morphine sold by the can or ounce; sometimes used to describe crude opium. |
| **Block Busters** | Barbiturate. |
| **Blond hash** | Golden brown-colored hashish. Less potent than Black Russian. |
| **Blow** | Cocaine, or to smoke marijuana in a "blow pot". |
| **Blow a fill** | Smoke opium. |
| **Blow a stick** | To smoke marijuana. |
| **Blow your mind** | Get high and lose mental control after smoking hallucinogens. |
| **Blow your mind roulette** | Very similar to "brown bag" roulette where the participants mix stimulant and depressant pills, then reach in and take a handful for ingestion. |
| **Blue** | A condition resulting from the overdose of a drug. |
| **Blues** | Barbiturates. |
| **Blue acid** | LSD. |
| **Blue Birds** | Barbiturates. |
| **Blue Cheer** | LSD. |
| **Blue Devils** | Barbiturates. |
| **Blue Heaven** | LSD. |
| **Blue Heavens** | Sodium amytal tablets or barbiturates. |

| | |
|---|---|
| **Blue Mist** | LSD. |
| **Blue Morning Glory seeds** | A hallucinogen. |
| **Blue Velvet** | Combination of elixir terpin hydrate, codeine, and tripelenamine. |
| **Body Drugs** | Opiates and barbiturates. |
| **Bogart** | To smoke a marijuana cigarette without sharing with others. |
| **Bogue** | Withdrawal sickness from drugs. |
| **BOL-148** | A congener of LSD. |
| **Bomb** | High potency drug or a prerolled marijuana cigarette resembling a king-size cigarette. |
| **Bombed Out** | High on drugs. |
| **Bomber** | Barbiturate. |
| **Bombita** | Cocaine in liquid form; or a vial of desoxyn, an amphetamine (stimulant); sometimes taken with heroin for a stronger high. |
| **Bong** | Pipe used for smoking marijuana or hashish. Normally used with water in the bowl; alcohol is sometimes added to the bowl instead of water and then drunk to increase the high of the marijuana or hashish. |
| **Boo** | Marijuana. |
| **Boogie** | To make a trip to smuggle in drugs. |
| **Boost** | Steal. |
| **Boot** | An autoerotic masturbatory-like experience; also, to feed blood back and forth into the "works", once the heroin is partially injected into the vein to obtain a more lasting effect. |
| **Boot and shoot** | A heroin addict who steals to support a habit. |
| **Booting** | Shooting up heroin but allowing it to back up into the syringe and then re-injecting it. |
| **Boss** | High grade drugs. Dominance. |
| **B.O.T** | Balance of time; sentence given to a parole violator. |
| **Bouncing powder** | Cocaine. |
| **Boy** | Heroin. |
| **Boxed** | High or intoxicated on drugs. |

| | |
|---|---|
| **Brain ticklers** | Barbiturate or amphetamine pills. |
| **Brick** | Compressed block of gum opium or morphine; also, a block of marijuana usually weighing 1 kilogram. |
| **Bring down** | To cause or lose a drug high. |
| **Broccoli** | Marijuana. |
| **Brody** | Pretending to be sick in order to obtain medication from a physician. |
| **Brother** | Heroin. |
| **Brown** | Brown heroin. Usually comes from Mexico. |
| **Brown bag roulette** | Individuals bring various types of drugs and throw them into a brown paper bag; then they reach in and take a handful and ingest them, sometimes with fatal results. Common among school children who go home and take whatever mom or dad were taking. |
| **Brown Rock** | 50% heroin used for smoking. |
| **Brownies** | Amphetamines. |
| **Brown Dots** | LSD. |
| **Brown Sugar** | Heroin. |
| **Brown Stuff** | Heroin, brown in color. |
| **Bull** | Policeman. |
| **Bumblebees** | Amphetamines. |
| **Bum bend** | Psychotic or panic reaction to LSD or STP. |
| **Bum kicks** | Troubled, worried, or depressed. |
| **Bum trip** | Bad experience or adverse reaction to a drug. |
| **Bundle** | A package of 25 five-dollar bags of heroin. |
| **Bunk habit** | Laying around an area that is used to smoke opium. Opium dens were where a person could inhale opium from those that were smoking without actually smoking it themselves. |
| **Burese** | Cocaine. |
| **Burn** | To cheat or to be cheated by someone taking money for drugs and not producing the goods. |
| **Burn artist** | One who uses the burn technique in order to cheat other addicts. |
| **Burned** | To be revealed as a drug peddler by a police informer; also an individual who has been cheated by a drug dealer. |

| | |
|---|---|
| **Burned out** | An addict who has kicked the habit; or when one is injecting the drugs, the veins can collapse or be "burned out." |
| **Bush** | Marijuana. |
| **Bust** | Police raid. |
| **Busted** | Arrested. |
| **Busters** | Barbiturates. |
| **Buttons** | Mescaline. |
| **Buzz** | A minor degree of euphoria or sensation of pleasure from a drug. |
| **Buzz, Rolling** | A moderate high that continues after the intake of drugs has ceased. |

## C

| | |
|---|---|
| **C** | Cocaine. |
| **Ca-ca** | Puerto Rican slang for counterfeit heroin. |
| **C.D.** | Glutethimide. |
| **Caballo** | Heroin. |
| **Cactus** | Mescaline. |
| **California Sunshine** | LSD. |
| **Canadian Black** | A type of marijuana grown in Canada. |
| **Canceled stick** | A cigarette filled with marijuana. |
| **Candy** | Cocaine. |
| **Cannabis** | Flowering or fruiting tops of the marijuana plant. |
| *Cannabis indica* | Hemp. |
| *Cannabis sativa* | A hemp plant which secretes a resin containing a hallucinogenic agent, THC. |
| **Canned** | To be arrested. |
| **Canned sativa** | Hashish. |
| **Canned stuff** | Packaged smoking opium. |
| **Cap** | Capsule of drugs. |
| **Cap-man** | Supplier of drugs. |
| **Carbona** | A brand of cleaning fluid of which the vapors are inhaled. |
| **Carbon tetrachloride** | Highly toxic and cancer-causing agent that has been sniffed for its deliriant effects. |

| | |
|---|---|
| **Carrying** | Possession of drugs on the person. |
| **Cartwheels** | Amphetamines. |
| **Cat** | Heroin. |
| **Catnip** | Sometimes sold to an unsuspecting buyer of marijuana. When burned, it gives off an odor very similar to marijuana. |
| **Cecil** | Cocaine. |
| **Chalk** | Amphetamines. |
| **Change** | Short jail time. |
| **Channel** | A vein for injecting heroin. |
| **Charged up** | Under the effects of the drug. |
| **Charas** | Unadulterated resin from marijuana plants. |
| **Charged** | Sudden, euphoric onset of opiates following injection. |
| **Chasing the bag** | Hustling for heroin. |
| **Chasing the dragon** | Smoking heroin (usually number 3 grade heroin). |
| **Chemical** | An illegally used drug. |
| **Chicharra** | Puerto Rican slang for a marijuana-tobacco cigarette. |
| **Chick** | Heroin; also a female. |
| **Chill** | Refusal to sell drugs to an addict who is suspected of being a police informant; also to kill someone. |
| **Chicken powder** | Amphetamines. |
| **Chillum** | A small clay pipe used to smoke ghanja. |
| **Chinese Red** | Heroin. |
| **Chinese White** | A very potent form of heroin. |
| **Chip** | To use heroin in small amount or occasionally. |
| **Chippy** | Potential addict, not quite yet hooked. |
| **Chiva** | Heroin. |
| **Chocolate chips** | LSD. |
| **Chloral hydrate** | A nonbarbiturate hypnotic and sedative drug. |
| **Christmas tree** | Green pill or green and white capsule, a look-alike drug containing caffeine and phenylpropanolamine sold on the street as speed. Barbiturates are also sometimes called "christmas trees" because of their bright colors. |

| | |
|---|---|
| **Cibas** | Glutethimide. |
| **Clean** | Not using drugs. |
| **Cleaning fluid** | Sniffed to get a high. |
| **Cleared up** | To withdraw from drugs or to be detoxified. |
| **Coasting** | Under the influence of drugs. |
| **Coffee** | LSD. |
| **Coast** | When a heroin addict starts to nod off. |
| **Coca** | The coca plant found in the hills of Java, Peru, and Bolivia; the coca leaf is used to make cocaine. |
| **Cocaine** | Cocaine hydrochloride, used both as a stimulant and anesthetic. |
| **Codeine** | Methyl morphine, an alkaloid of the opium family. |
| **Cohoba** | A hallucinogenic snuff made by the natives of Trinidad. |
| **Coke** | Cocaine. |
| **Coke Head** | One who uses cocaine. |
| **Coked-up** | Under the influence of cocaine. |
| **Colas** | The flowering top of the marijuana plant. |
| **Cold** | Same as cold turkey; to quit using drugs abruptly. |
| **Cold bust** | An arrest made when the person is concealing drugs. |
| **Cold shot** | A bad drug deal. |
| **Cold turkey** | An abrupt withdrawal from drugs not generally under the supervision of a physician or use of controlling medication. |
| **Collapsed veins** | Veins that swollen, blocked, or thrombosed. The veins cannot be used to inject drugs. |
| **Collar** | To make an arrest. |
| **Colombina** | A very high grade of marijuana. |
| **Come down** | To start to feel sick as effects of drug wear off. |
| **Conar** | A type of cough syrup containing any nonaddicting alkaloids of opium. |
| **Connect or cop a fix** | To buy drugs or find a source of drugs. |
| **Connection** | Street peddler of drugs; source of supply. |
| **Connection dough** | Money used to make a purchase from a connection. |

| | |
|---|---|
| **Contact lens** | LSD. |
| **Cook** | To make a mixture of heroin and water, then heat it in a bottle cap or spoon; the mixture is then injected. |
| **Cook up a pill** | To smoke opium. |
| **Cooker** | Bottle cap or spoon used for dissolving heroin in water over a flame. |
| **Cooker** | A person who produces illegal drugs in a clandestine laboratory. |
| **Cooking** | The act of preparing illicit drugs in a clandestine drug lab. |
| **Cool** | Trustworthy or one who is careful when making a buy of drugs. |
| **Cop** | To buy drugs or comparing cost of various types of drugs. |
| **Cope** | To carry on activities of daily life effectively while under the influence of drugs. |
| **Copilot** | A person who sits with someone when an individual takes LSD. |
| **Copilots** | Amphetamines; sometimes used for Dexedrine tablets. |
| **Cop Man** | Middle man or pusher. |
| **Cop sickness** | Anxiety or discomfort experienced by an addict before his or her next fix or injection. |
| **Cotton** | A piece of cotton ball through which the cooked heroin is pulled in order to remove some of the impurities. |
| **Cotton habit** | An irregular habit of using drugs. |
| **Count** | The number or weight of drugs in a sale. |
| **Courage pills** | Amphetamines. |
| **Crack** | Concentrated form of cocaine. |
| **Crack head** | One who uses crack. |
| **Crack house** | Apartment, room, or house where crack is used; usually fortified to keep out law enforcement. |
| **Crap** | Heroin that has been "stepped on" or diluted. |
| **Crank** | Amphetamines. |
| **Crank bugs** | Hallucinations following a heavy dose of amphetamines. |

| | |
|---|---|
| **Crash** | To fall asleep while using drugs; to come down hard and fast from a high or trip. |
| **Crankster gangster** | Name for the "cooker" or producer of illegal drugs in a clandestine laboratory. |
| **Crater** | Large hole in the flesh over a vein that has been used repeatedly to inject drugs. |
| **Creep** | An addict who begs to get his drugs or tries other people's drugs. |
| **Crib** | Home or apartment; sometimes referred to as a "shooting gallery", a place where drugs are shot up. |
| **Croaker** | An unscrupulous physician who sells drugs or prescriptions to illicit drug users. |
| **Crossroads** | Amphetamines. |
| **Crosstop** | Small circular white pill with "X", a look-alike drug containing caffeine and phenylpropanolamine, sold on the street as speed. |
| **Crumbs** | Small change or small amounts of drugs. |
| **Crutch** | Device used for holding shortened butt of marijuana cigarette; also referred to as a "roach" or "clip". |
| **Crystal** | Speed or methamphetamine; can also refer to cocaine. |
| **Crystal Palace** | A place where speed users get together to do drugs. |
| **Cube** | Morphine. |
| **Cubes** | LSD. |
| **Cura** | From Spanish, a shot of heroin. |
| **Cure** | A length of stay in a hospital. |
| **Cut** | To adulterate drugs, as with lactose (milk sugar) mannitol, or quinine. |
| **Cut ounces** | To further dilute the drug. |
| **Cutting** | The process of diluting heroin or other drugs with various chemicals. |
| **Cupcakes** | LSD. |

### D

| | |
|---|---|
| **Dabbling** | Irregular use of narcotics. |
| **Dead on arrival (DOA)** | Phencyclidine. |
| **Deadly nightshade** | Belladonna. |

| | |
|---|---|
| **Deal** | To sell drugs. |
| **Dealer** | Drug pusher. |
| **Dealing** | Selling drugs. |
| **Deck** | A glassine bag filled with drugs. |
| **Dexie** | Dexadrine or amphetamine; may also be a look-alike drug. |
| **Dex** | Amphetamine. |
| **Dinosaur shorts** | Name for foul smelling drugs, generally as they are produced in a clandestine laboratory. |
| **Dill** | Dill weed, crushed and smoked. |
| **Dime bag** | Ten dollar ($10.00) purchase of drugs. |
| **Dip and dab** | To try heroin occasionally. |
| **Dirty** | To be caught with drugs. |
| **Ditch weed** | Low grade marijuana, generally found growing along the highways in the West and Midwest. |
| **DMT** | Hallucinogen. |
| **Do** | Using drugs. |
| **D.O.E.** | Taking drugs. |
| **Domestic** | Marijuana grown in the U.S. |
| **Doobie** | Marijuana cigarette. |
| **Doojee** | Heroin. |
| **Dope** | Any illicit drug. |
| **Dope fiend** | A term used by addicts to describe themselves. |
| **Dope run** | A trip to purchase drugs. |
| **Double cross** | Amphetamines. |
| **Double trouble** | A mixture of barbiturates and amphetamines. |
| **Double uoglobe** | Highly refined heroin. |
| **Do up** | To inject heroin. |
| **Down** | Using downers, barbiturates. |
| **Downers** | Tranquilizers. |
| **Down trip** | Boring or to bring down off of a high. |
| **Downs** | Barbiturates. |
| **Drag** | A bad trip or a bum trip. |
| **Dragged** | A frightful experience under the influence of a hallucinogen. |

| | |
|---|---|
| **Dream** | Cocaine. |
| **Dreamer** | Morphine. |
| **Dried out** | Detoxified. |
| **Dripper** | A dropper used to inject drugs; mainly to inject heroin. |
| **Drop** | To swallow a pill. |
| **Drop It** | To conceal drugs. |
| **Dry up** | To temporarily stop using drugs. |
| **Dry high** | Marijuana. |
| **Duby** | Marijuana. |
| **Duji** | Heroin. |
| **Dummies** | Any phony or counterfeit drug, often intended to appear to be heroin; usually a substance like sugar, talcum powder, kitchen cleanser, etc. |
| **Dust** | To sell drugs; sometimes for cocaine. |
| **Dusting** | Mixing heroin with marijuana. |
| **Dynamite** | Drugs of exceptional purity and strength. |
| **Dyno** | A high-potency heroin. |

## E

| | |
|---|---|
| **Eater** | A drug user who ingests his drugs. |
| **Eight oz.** | 1/8 ounce of marijuana. |
| **Electric** | Containing a psychedelic drug. |
| **Electric Kool-Aide** | A punch containing LSD. |
| **Elephant** | Phencyclidine, angel dust or PCP. |
| **Embroidery** | Scars left from injecting into the veins leaving a pattern. |
| **Eye dropper** | Medicine dropper used with hypodermic needle as makeshift syringe for injecting drugs. |
| **Eye opener** | Amphetamine; also the first narcotics shot of the day. |

## F

| | |
|---|---|
| **Factory** | A clandestine laboratory producing drugs. |
| **Fake a blast** | To simulate taking of drugs; very risky ploy sometimes used by undercover officers to convince drug users that they are actually taking drugs. |
| **Falling out** | To doze while under the influence of drugs. |

| | |
|---|---|
| **Fatty** | A very thick marijuana cigarette. |
| **Fed** | A member of a federal law enforcement group like the DEA or FBI. |
| **Feeling the monkey coming on** | First signs of withdrawal. |
| **Fiend** | An addict who uses large amounts of drugs. |
| **Fingers** | Hashish. |
| **Firing the ack-ack gun** | A way of taking heroin, putting the heroin on the tip of a burning cigarette. |
| **Fish** | A patient newly admitted to hospital. |
| **First line** | Morphine. |
| **Fit** | Equipment used to inject drugs. |
| **Fix** | A shot of narcotics or drugs. |
| **Flake** | Cocaine. |
| **Flaky** | A bit crazy, or the flaky consistency of cocaine. |
| **Flash** | A sense of colors or a rush of pleasurable feelings. |
| **Flashing** | Glue sniffing. |
| **Flea powder** | Poor or inferior grade of heroin. |
| **Flip out** | Panic, temporary, or chronic psychotic reaction to drugs. |
| **Floating** | High on drugs. |
| **Flow** | To let the effects of the drugs take over. |
| **Flowers** | Flowering tops of marijuana plants. |
| **Fluff** | Cotton used to filter heroin. |
| **Flunky** | A fool, or a person who takes great risks in using drugs. |
| **Flush** | Blushing or hot feeling of the skin. |
| **Fly agaric** | A poisonous mushroom. |
| **Flying** | A state of drug intoxication. |
| **Flying saucers** | A variety of morning glory seeds which have hallucinogenic properties. |
| **Fold up** | To cease selling or using drugs. |
| **Fool** | Similar to flunky. |
| **Footballs** | Opiates. |
| **Forwards** | Amphetamines |

| | |
|---|---|
| **Fourth oz.** | 1/4 ounce of marijuana. |
| **Freak** | Bizarre, weird, freakish. |
| **Freaking out** | Panic reaction to drugs. |
| **Freak up** | A kind of hoax, making people believe your behavior is drug-induced. |
| **Freaky** | Bizarre. |
| **Freeze** | To refuse to sell drugs. |
| **Front** | Putting money up "front" for a drug deal, or as putting on a "front" to impress people. |
| **Fruit salad** | A game played where individuals take one pill from every bottle found in the house and ingest them. |
| **Full moon** | A large peyote chunk usually about 4 inches in diameter or the whole top of a peyote cactus. |
| **Fun** | A unit of measure of opium; one dose. |
| **Fuzz** | Police or law enforcement personnel. |

### G

| | |
|---|---|
| **Gage** | Marijuana. |
| **Ganga** | Marijuana. |
| **Gangster** | Marijuana |
| **Gangster pills** | Barbiturates or other sedative pills. |
| **Gaping** | Yawning, an early withdrawal symptom of opiates. |
| **Garbage** | Dilute heroin. |
| **Garbage Head** | One who takes any drug to get high. |
| **Gasoline** | Sniffing the fumes from gasoline. |
| **Gauge** | A marijuana cigarette. |
| **Gay** | A person who takes drugs to get the exhilarating effects; also, a homosexual. |
| **G.B.** | A barbiturate pill. |
| **Gee Head** | An addict who takes paregoric. |
| **Geronimo** | An alcoholic beverage often containing wine mixed with drugs. |
| **Get down** | To inject heroin. |
| **Get in the groove** | Familiar with drug trafficking or taking drugs. |
| **Get off** | To inject heroin. |
| **Get on** | Taking drugs for the first time. |

| | |
|---|---|
| **Get one's yen off** | To satisfy one's drug needs. |
| **Get the habit off** | Taking drugs at a given time. |
| **Getting on** | Smoking marijuana. |
| **Ghost** | LSD. |
| **Giggle weed** | Marijuana. |
| **Gimmicks** | Equipment for injecting drugs. |
| **Girl** | Cocaine. |
| **Globetrotter** | An addict who circulates around the drug circuit seeking the best buy in drugs. |
| **Glow** | High or a feeling of euphoria. |
| **Glue sniffing** | Inhaling toluene or other model glue. |
| **God's medicine** | Morphine. |
| **Going down** | Everything is going well for the addict. |
| **Going high** | A pleasurable, continuing state of drug intoxication. |
| **Gold** | High potency marijuana. |
| **Gold Dust** | Cocaine. |
| **Golden Leaf** | Marijuana, generally Acapulco Gold. |
| **Goods** | Narcotics. |
| **Good sick** | Nausea and vomiting associated with using heroin or other opiates; addicts do not consider this an unfavorable experience following injection of a drug. |
| **Good stuff** | Heroin of superior or excellent quality. |
| **Goofballs** | Barbiturates. |
| **Goofers** | Barbiturates. |
| **Goofing** | Under the influence of barbiturates. |
| **Gorilla pills** | Barbiturates. |
| **Grass brownies** | Brownies made with marijuana in them. |
| **Grass** | Marijuana. |
| **Gravy** | Mixture of blood from the addict's vein and heroin that is reheated in the cooker and cannot be reinjected. |
| **Greasy junkie** | A passive addict. |
| **Green** | A very cheap and low grade marijuana. |
| **Green Dragons** | Barbiturates. |

| | |
|---|---|
| **Greenies** | Amphetamines. |
| **Grefa** | Marijuana. |
| **Greta** | Marijuana. |
| **Grifa** | Marijuana. |
| **Griffo** | Marijuana. |
| **Groovers** | Teenagers who use drugs for the thrills, e.g., sexual and bizarre mental experiences. |
| **Grooving** | Becoming intoxicated by drugs. |
| **Groovy** | A high experienced on drugs. |
| **Ground control** | A person who attends a person going on an LSD trip. |
| **Grower** | Marijuana cultivator, uninvolved in smuggling of drugs. |
| **Guide** | An experienced LSD user. |
| **Gum** | Opium in the raw used for smoking. |
| **Gun** | Hypodermic needle. |
| **Guru** | Experienced LSD user who guides new users. |
| **Gutter** | Veins inside the elbow; heroin is injected into these veins. |

## H

| | |
|---|---|
| **H** | Heroin. |
| **Hay** | Marijuana. |
| **Hard drugs** | Narcotics. |
| **Hard stuff** | Opiates. |
| **Harry** | Heroin. |
| **Hash** | Hashish. |
| **Hassle** | Annoyances, bothering, exertion in obtaining drugs. |
| **Hassling** | Buying drugs. |
| **Hawk, The** | LSD. |
| **Hay** | Marijuana. |
| **H-caps** | Heroin sold in gelatin capsules. |
| **Haze** | LSD. |
| **Head** | A drug user. |
| **Head drugs** | Any drug that affects the mind. |

| | |
|---|---|
| **Hearts** | Amphetamines. |
| **Hearts** | Look-alike drugs containing caffeine and phenyl-propanolamine sold on the street as speed; may be pink, orange, white, or tan in shape of a heart. |
| **Heat, The** | Law enforcement. |
| **Heaven dust** | Cocaine. |
| **Heavenly Blue** | A variety of morning glory seeds. |
| **Heavy drugs** | Narcotics. |
| **Heeled** | Carrying a weapon. |
| **Hemp** | Marijuana. |
| **Her** | Cocaine. |
| **Herb** | Marijuana. |
| **High** | Under the influence of drugs. |
| **Him** | Heroin. |
| **Hit** | Dose of drugs, or drag on a marijuana cigarette. |
| **Hocus** | Heroin; also used for morphine. |
| **Hog** | Phencyclidine. |
| **Homegrown** | Marijuana grown in the U.S. |
| **Hooked** | Dependent upon drugs. |
| **Hop** | Opium. |
| **Hop dog** | An opium addict. |
| **Horn** | To sniff or inhale a drug. |
| **Horror drug** | A drug containing belladonna. |
| **Horrors** | A bad experience with drugs. |
| **Horse** | Heroin |
| **Hot bust** | An arrest made from information from another user. |
| **Hot shot** | An often fatal injection of a drug that looks like heroin but is often a poison. |

### I

| | |
|---|---|
| **Ice cream habit** | Irregular or modest use of drugs. |
| **Idiot pill** | Barbiturates. |
| **Incentive** | Cocaine. |
| **Indian hay** | Marijuana. |
| **Indian hemp** | Marijuana. |

| | |
|---|---|
| **Inhalants** | Deliriants. |
| **Into** | As "into" using drugs. |

## J

| | |
|---|---|
| **J** | Marijuana. |
| **Jab** | To inject drugs. |
| **Jay** | Marijuana. |
| **Jam** | Cocaine. |
| **Jammed up** | To have taken an overdose. |
| **Jane** | Marijuana. |
| **Jelly beans** | Amphetamines. |
| **Johnson grass** | A type of low potency marijuana found in Texas. Named for Lyndon B. Johnson. |
| **Joint** | Prison; also a marijuana cigarette. |
| **Joy juice** | Chloral hydrate. |
| **Joy pop** | To inject a small amount of heroin. |
| **Juice** | Alcohol. |
| **Juice head** | One who uses liquor. |
| **Junk** | Heroin. |
| **Junkie** | Heroin addict. |

## K

| | |
|---|---|
| **Key** | A kilo (2.2 lbs) of any drug, especially marijuana or hashish compressed into a "brick". Short for kilogram (kilo). |
| **Ki** | A kilogram. |
| **Kick a habit** | To be detoxified from drugs. |
| **Kick cold turkey** | To undergo withdrawal. |
| **Kicking** | Ridding the body of drugs. |
| **Kick sticks** | Marijuana. |
| **Kick the habit** | To rid oneself of drugs. |
| **Kif** | Hashish. |
| **Killer** | Very good but potent drugs. |
| **Killer weed** | Phencyclidine, when combined with marijuana or other plant material. |
| **Kilter** | Marijuana. |

| | |
|---|---|
| **King Kong** | A very large and expensive heroin habit. |
| **King Kong pills** | Barbiturates. |
| **Kit** | Equipment for injecting drugs, especially heroin. |
| **Knock out drops** | Chloral hydrate. |

## L

| | |
|---|---|
| **L** | LSD. |
| **Lady** | Cocaine. |
| **Lady Snow** | Cocaine. |
| **Lame** | A conventional person, a square. |
| **Laudanum** | Alcohol and opium mixture. |
| **Laughing gas** | Nitrous oxide. |
| **Layout** | Equipment for injecting heroin. |
| **Lay up** | To stockpile drugs. |
| **Leapers** | Amphetamines. |
| **Lemon** | Worthless powder sold as heroin. |
| **Lid** | The size of a standard marijuana transaction, generally about an ounce. |
| **Lid poppers** | Amphetamines. |
| **Light stuff** | Marijuana. |
| **Lightning** | Amphetamines. |
| **Lit up** | Under the influence of a drug. |
| **Load** | Bulk sale of heroin, thirty $3.00 bags; (half load is fifteen $3.00 bags); also a supply of drugs, sometimes referred to as a "stash". |
| **Loco** | Marijuana; also called loco weed. |
| **Love drug** | MDA, a synthetic drug similar to amphetamine and mescaline. |
| **Lude** | Methaqualone, a depressant drug. |
| **Luding out** | Using methaqualone. |
| **Lumber** | Waste from marijuana. |
| **Lush** | A heavy drinker. |

## M

| | |
|---|---|
| **M** | Morphine. |
| **Morph** | Morphine. |

| | |
|---|---|
| **Magic mushroom** | Psilocybin. |
| **Mainline** | Injecting heroin directly into a vein. |
| **Main stash** | Where drugs are hidden. |
| **Manicure** | To clean and prepare marijuana by removing seeds and stems so that the marijuana can be rolled into cigarettes. |
| **MDA** | "Mellow Drug of America", synthetic drug similar to mescaline and amphetamine. |
| **Marshmallow Reds** | Barbiturates. |
| **Mary Jane** | Marijuana. |
| **Matchbox** | Usually five to eight joints of marijuana. |
| **Miss Emma** | Morphine. |
| **Mellow Yellows** | LSD. |
| **Merchandise** | Drugs. |
| **Mesc** | Mescaline. |
| **Mescal** | Mescaline. |
| **Mescal buttons** | Mescaline. |
| **Meth** | Amphetamines. |
| **Meth freak** | A frequent, habitual user of methedrine. |
| **Mexican Brown** | A grade of Mexican marijuana. |
| **Mexican Green** | Low resin, commonest of Mexican marijuana. |
| **Mexican Locoweed** | Plant found in southwestern U.S., which causes intoxication in animals. |
| **Mexican mud** | Heroin. |
| **Mexican mushrooms** | Hallucinogenic mushrooms. |
| **Mexican Reds** | Barbiturates. |
| **Mickey Finn** | An alcoholic drink containing knock-out drops of chloral hydrate. |
| **Microdot** | Small round pill of LSD. May be purple, red, or orange. May be on transfers or decals that have cartoon characters or animals. |
| **Milk sugar** | Crystals of lactose, closely imitating heroin. Often used to cut or "step on" drugs. |
| **Minibennies** | Amphetamines. |
| **Monkey** | An addict's dependency on drugs; also a feeling that there is a separate person living within the body. |

| | |
|---|---|
| **Moon** | Mescaline. |
| **Mooters** | Marijuana cigarettes. |
| **Mootie** | Marijuana. |
| **Morning glory seeds** | Seeds of the bindweed family of plants that have a hallucinogenic effect. (Seeds sold in garden shops now have a coating that makes the user vomit. Seeds contain a natural form of LSD.) |
| **Morning shot** | First fix of the day. |
| **Morf** | Morphine. |
| **Morpho** | Morphine. |
| **Morphy** | Morphine. |
| **Mota** | Marijuana. |
| **Mother** | A pusher or dealer in drugs. |
| **Mud** | Heroin, can also be morphine. |
| **Mushroom** | Psilocybin. |
| **Mutah** | Marijuana. |

## N

| | |
|---|---|
| **Nab** | To be arrested. |
| **Nail** | Hypodermic syringe. |
| **Nail polish remover** | Acetone, sniffed to get intoxicating effects. |
| **Nailed** | To be arrested. |
| **Narc or narco** | Narcotics officer. |
| **Nark** | Narcotics officer. |
| **Nebbies** | Barbiturates. |
| **Needle** | Hypodermic needle. |
| **Needle freak** | One who enjoys the pain of using a hypodermic needle. |
| **Negative** | A bad trip, a bummer. |
| **Nemmies** | Nembutal capsules. |
| **Nickel bag** | $5.00 bag of drugs. |
| **Nimbies** | Barbiturates. |
| **Nitrous oxide** | Anesthetic gas, an inhalant anesthetic. |
| **Nixon** | Low potency narcotics. |
| **Nod** | Nodding, a drowsy, dreamy state. |

| | |
|---|---|
| **Nod on** | The sleep-like state, a result of using drugs, usually in reference to heroin. |
| **Nose candy** | Cocaine. |
| **Nuggets** | Amphetamines. |
| **Number** | A marijuana cigarette. |
| **Number 3 grade heroin** | Heroin used for smoking. |
| **Number 4 grade heroin** | Highly refined heroin, used for injection. |
| **Nutmeg** | Dried seeds of the East Indian evergreen used for seasoning. Taken in powdered form, has mild mind-altering properties. |

## O

| | |
|---|---|
| **O** | Opium in the raw form. |
| **OP** | Opium. |
| **O.D** | Overdose. |
| **Off** | To kill or to be high on drugs. |
| **Oil** | Oil from hashish, very high in THC. |
| **On** | Using drugs. |
| **On the needle** | Using intravenous drugs. |
| **Opiate** | A natural or semisynthetic drug derived from opium. |
| **Oranges** | Amphetamines. |
| **Orange mushrooms** | LSD. |
| **Orange wedges** | LSD. |
| **Outfit** | Equipment used for injecting drugs. |
| **Overjolt** | Overdose of a drug. |
| **Owsley** | LSD. |
| **Oz** | One ounce of a drug. |

## P

| | |
|---|---|
| **P** | Peyote. |
| **Packed up** | High on drugs. |
| **Panama** | Panama gold; marijuana. |
| **Panama Gold** | Marijuana. |
| **Panama Red** | Marijuana. |
| **Panic Man** | An addict who can not obtain drugs. |

| | |
|---|---|
| **Paper acid** | LSD. |
| **Paper bag** | Container of drugs. |
| **Papers** | Any thin paper used to roll marijuana cigarettes. |
| **Paradise** | Cocaine. |
| **PCP** | Phencyclidine. |
| **Peace pill** | Phencyclidine. |
| **Peaking** | Highest stage of intensity from an LSD trip. |
| **Pearls** | Amyl nitrate. |
| **Peanuts** | Barbiturates. |
| **Peddler** | Supplier of drugs. |
| **Pellets** | Capsules containing LSD. |
| **Pep pills** | Amphetamines. |
| **Per** | Prescription. |
| **Peter** | Chloral hydrate, a sedative. |
| **P.D.K.** | Look-alike drug containing caffeine and phenyl-propanolamine sold on the street as speed. |
| **Phennies** | Opiates. |
| **Piece** | A weapon, usually a pistol; also an ounce of heroin. |
| **Pillhead** | An addict on pills. |
| **Pin** | Marijuana cigarette. |
| **Pinhead** | Thinly rolled marijuana cigarette. |
| **Pinks** | Barbiturate. |
| **Pink Heart** | Look-alike drug containing caffeine and phenyl-propanolamine sold in shape of a heart or football. |
| **Pink Ladies** | Barbiturates. |
| **Pinks** | Barbiturates. |
| **Pod** | Marijuana. |
| **Point** | Hypodermic needle. |
| **Poison** | Heroin, cocaine. |
| **Poison Act** | Federal narcotics act. |
| **Poison People** | Heroin addicts. |
| **Poke** | A puff on a marijuana cigarette. |
| **Pop** | To inject drugs; also to be arrested. |
| **Popped** | To be arrested. |

| | |
|---|---|
| **Poppers** | Amyl nitrate. |
| **Popping** | Skin injection where the addict injects a small amount of drugs just under the skin; also, to swallow a pill. |
| **Pot** | Marijuana; also called "grass", "joints", "weed", "tea", "hay", "reefer", "gauge", "maggies", "Mary Jane", etc. |
| **Potlikker** | A tea made from the seeds and stems of the marijuana plant after it has been "manicured". |
| **Poppy** | Opium. |
| **Powder** | Heroin. |
| **Powdering off** | Preparing drugs in a clandestine laboratory. |
| **P.R.** | Panama red, marijuana. |
| **Primo** | From Spanish, number one quality. |
| **Puff** | To smoke opium. |
| **Pure** | Undiluted heroin. |
| **Purple Hearts** | Phenobarbital. |
| **Push** | To deal in drugs. |
| **Pusher** | Drug dealer or peddler. |
| **Pushing** | To be selling drugs. |
| **Put somebody on** | To give a person a marijuana cigarette to smoke. |

## Q

| | |
|---|---|
| **Quarter moon** | Hashish. |
| **Quas** | Methaqualone. |
| **Quads** | Methaqualone. |
| **Quill** | Matchbook cover used to inhale drugs; the powdered drug is placed in the fold of the matchbook. |
| **Quieters** | Tranquilizers. |

## R

| | |
|---|---|
| **Ragweed** | Low grade marijuana. |
| **Rainbows** | Tuinal capsules, a barbiturate. |
| **Rainy Day Woman** | Marijuana cigarette. |
| **Rap** | To talk. |
| **Red** | Barbiturates; also Panama red (marijuana). |

| | |
|---|---|
| **Red birds** | Seconal capsules, a barbiturate. |
| **Reds and blues** | Barbiturates. |
| **Red bulls** | Seconal capsules, a barbiturate. |
| **Red chicken** | Chinese heroin. |
| **Red devils** | Seconal capsules, a barbiturate. |
| **Red dirt marijuana** | Wild growing marijuana. |
| **Red rock** | Chinese heroin in granular form. |
| **Reds** | Seconal capsules, a barbiturate. |
| **Reefer** | Marijuana cigarette. |
| **Rig** | Equipment for injecting drugs. |
| **Righteous bush** | Marijuana. |
| **Ripped** | Intoxicated by drugs. |
| **Rippers** | Amphetamines. |
| **Roach** | Butt end of a marijuana cigarette. |
| **Roach clip** | Instrument used to hold a marijuana cigarette being smoked; may be an alligator clip or specially designed devices. |
| **Robin's egg** | Look-alike drug containing caffeine and phenylpropanolamine sold as speed; usually blue or white with blue specks. |
| **Rock** | Cocaine, or heroin. |
| **Roll** | A wrapped quantity of drugs. |
| **Root** | A marijuana cigarette. |
| **Rope** | Hashish. |
| **Roses** | Amphetamines. |
| **Royal Blue** | LSD. |
| **Rush** | Warm or ecstatic feeling throughout the body that many addicts report 15 to 30 seconds after intravenous injection of heroin (often as a tingling sensation in the abdomen or genital area.) |

## S

| | |
|---|---|
| **Sacrament** | LSD. |
| **Sacred mushrooms** | Mexican mushrooms, psilocybin. |
| **Sativa** | Marijuana. |
| **Scag** | Heroin. |

| | |
|---|---|
| **Schmack** | Heroin. |
| **Schmeck** | Heroin. |
| **Schmecker** | A heroin user. |
| **School boy** | Codeine. |
| **Scoop** | Matchbook cover used to sniff heroin or cocaine. |
| **Score** | To buy drugs. |
| **Scoring** | Making a purchase. |
| **Script** | Prescription. |
| **Set up** | To be positioned by an addict or undercover agent for an arrest. |
| **Sherman** | Marijuana cigarette laced with PCP. |
| **Shit** | Heroin. |
| **Shlook** | To puff on a marijuana cigarette. |
| **Shooting gallery** | Place, such as a home, room, or apartment where addicts inject drugs (normally heroin). |
| **Shooting up** | Injecting drugs. |
| **Shot** | An injection of drugs. |
| **Short count** | A short weight, volume, or count of drugs. |
| **Shove** | To sell drugs at the street level. |
| **Shroom** | Psilocybin; magic mushroom; a hallucinogen. |
| **Sick** | Suffering from withdrawal symptoms. |
| **Skagg** | Heroin. |
| **Skin** | A paper for rolling marijuana cigarettes. |
| **Skin popping** | Intramuscular or subcutaneous injection of heroin. |
| **Skin shot** | Subcutaneous injection of drugs. |
| **Sleepers** | Barbiturates. |
| **Sleeping pills** | Barbiturates. |
| **Sleep walker** | Heroin addict. |
| **Smack** | Heroin. |
| **Smashed** | Intoxicated on alcohol or drugs |
| **Smoke** | Marijuana. |
| **Snappers** | Amyl nitrate vials. |
| **Sniff** | Cocaine. |

| | |
|---|---|
| **Sniffing** | Inhaling drugs through the nose, often done with cocaine. |
| **Snop** | Marijuana or a marijuana cigarette. |
| **Snorting** | Inhaling drugs through the nose, often done with cocaine. |
| **Snort** | Cocaine. |
| **Snow** | Cocaine. |
| **Soapers** | Methaqualone. |
| **Soft drugs** | Nonnarcotic drugs, such as marijuana, amphetamines, barbiturates, and some minor hallucinogens. |
| **Sopes** | Methaqualone. |
| **Soles** | Hashish. |
| **Source, The** | Supplier of drugs. |
| **Spaced** | A dislocation of time or reality usually from hallucinogens. |
| **Spaced out** | Spaced; removed from reality. |
| **Speed** | Amphetamine or "uppers". |
| **Speed ball** | Heroin mixed with cocaine. |
| **Speed freak** | A heavy intravenous user of methamphetamines. |
| **Spike** | Hypodermic needle. |
| **Splash** | Amphetamines. |
| **Split** | To run away. |
| **Spoon** | Sixteenth of an ounce of heroin or cocaine; some addicts wear a tiny spoon as a piece of jewelry. |
| **Spring** | To treat someone to a marijuana cigarette. |
| **Square** | Someone who does not do drugs. |
| **Stamp** | Small square paper stamp containing LSD; usually has some type of figure or decal on the paper. |
| **Stardust** | Cocaine. |
| **Stash** | A hidden cache of drugs. |
| **Stepped on** | Cut, diluted. |
| **Stick** | Marijuana cigarette. |
| **Stoned** | Usually high, primarily on LSD or other hallucinogen. |

| | |
|---|---|
| **Stool** | From "stool pigeon", one who gives the police information. |
| **STP** | A hallucinogen. |
| **Straight** | An addict's feeling of well-being after taking drugs; also, going without drugs. |
| **Strawberry Fields** | LSD. |
| **Street, on the** | Using drugs. |
| **Strung-out** | To be addicted; more specifically, to be badly in need of a "fix". |
| **Stuff** | Heroin or any other drug. |
| **Stumblers** | Barbiturates. |
| **Sugar** | LSD. |
| **Sugar weed** | Marijuana, soaked in a solution of sugar and water. |
| **Sunshine** | LSD. |
| **Swing Man** | Pusher or dealer in drugs. |
| **System, The** | The body's tolerance to drugs. |

## T

| | |
|---|---|
| **Tabs** | Capsules containing LSD. |
| **Taken off** | To be robbed. |
| **Take off** | To get high. |
| **Taking off** | To be robbed. |
| **Take the cure** | Go to the hospital. |
| **Taking a trip** | Under the effects of drugs. |
| **Tapita** | From Spanish, a bottle cap used to cook heroin. |
| **Tar** | Opium. |
| **Tea** | Marijuana |
| **Tea bag** | Smoking marijuana. |
| **Tea head** | A frequent heavy user of marijuana. |
| **Ten-cent pistol** | A bag of supposed heroin actually containing poison. |
| **Texas tea** | Marijuana. |
| **THC** | Tetrahydrocannabinol, the psychoactive ingredient in marijuana (cannabis). |

| | |
|---|---|
| **The Man** | Policeman. |
| **Three-fifty-seven magnum** | From 0.357 magnum, a bullet-shaped pink and white look-alike containing caffeine and phenylpropanolamine sold on the street as speed. |
| **Thing** | Heroin. |
| **Throwing rocks** | Committing a violent crime. |
| **Thrusters** | Amphetamines. |
| **Thumb** | A fat marijuana cigarette. |
| **Ticket** | LSD. |
| **Tie** | A belt or tourniquet used to distend veins so that they can be injected into. |
| **Tie off** | Tie (distend) the veins. |
| **Tie up** | To tie the veins for injection. |
| **Tighten up** | To give someone drugs. |
| **Tin** | A can of smoking opium. |
| **Tingle** | The rush or onset of a drug's reaction. |
| **TJ** | Mexican marijuana, from Tijuana, Mexico. |
| **Toak (or toke)** | A puff on a marijuana cigarette. |
| **Toak (or toke) pipes** | Short-stemmed pipes in which marijuana or hashish is smoked. |
| **Toke up** | To smoke marijuana. |
| **Tools** | Equipment for using drugs. |
| **Top** | A wheat-straw cigarette paper, used to smoke marijuana. |
| **Torn up** | Intoxicated by drugs. |
| **Tossed** | Searched for drugs. |
| **Toy** | A small box of opium. |
| **Tracks** | Veins collapsed by constant injection; also needle scars from injections. |
| **Tracking** | Scars left from constant injections. |
| **Trap** | A hiding place for drugs. |
| **Travel agent** | A user or supplier of LSD. |
| **Trip** | The LSD or hallucinogenic drug experience. |
| **Trips** | LSD. |
| **Truck drivers** | Amphetamines. |

| | |
|---|---|
| **Turkey** | Bag of nonnarcotic chemicals sold as a narcotic. |
| **Turn abouts** | Amphetamines. |
| **Turn on** | To introduce someone to drugs and give them their first drug experience; also to excite, usually sexually. |
| **Turned off** | Withdrawn from drugs. |
| **Turned out** | Introduced to the drug life. |
| **Turp** | Terpin hydrate, a medication found in some cough syrups. |
| **Twist** | Marijuana. |
| **Twisted** | Suffering withdrawal symptoms. |
| **Tying up** | Applying a tie or tourniquet in order to raise the veins for injection. |

## U

| | |
|---|---|
| **U.C.** | Undercover narcotics agent. |
| **Uncle** | Uncle Sam, a narcotics agent. |
| **Up-tight** | Feel ill, desperate, in desire, need; anxious. |
| **Ups or uppers** | A term for amphetamines. |
| **Using** | Taking drugs. |

## V

| | |
|---|---|
| **Voyager** | A person under the influence of LSD. |

## W

| | |
|---|---|
| **Wake-ups** | Amphetamines; also the first injection of the day that a drug addict takes. |
| **Wasted** | Heavily under the influence of a drug. |
| **Water** | Amphetamines. |
| **Water pipe** | "Bong", or pipe used to smoke marijuana. |
| **Wedges** | LSD. |
| **Weed** | Marijuana. |
| **Weeds** | Hashish or pot. |
| **Weekend habit** | Small, irregular drug habit. |
| **Weekend warrior** | An irregular drug user. |
| **Weight** | Large quantity of drugs. |
| **Whacked** | Cut or adulterated drugs. |

| | |
|---|---|
| **White** | Cocaine. |
| **White Lady** | Cocaine. |
| **White Lightning** | LSD; also illegal whiskey. |
| **White stuff** | Morphine. |
| **Whites** | Benzedrine pills. |
| **Window pane** | LSD in small squares. |
| **Wings** | The first mainline shot an addict ever takes, as in "earning their wings." |
| **Wiped out** | Acutely intoxicated from a drug. |
| **Wired** | Addicted to heroin; also high on amphetamines. |
| **Works** | Hypodermic needle. |
| **Working** | Obtaining money for drugs. |

## Y

| | |
|---|---|
| **Yellow** | Barbiturates. |
| **Yellows** | LSD. |
| **Yellow Jackets** | Nembutal capsules; barbiturates. |
| **Yen shee** | Opium ash. |
| **Yen shee suey** | Opium wine. |

## Z

| | |
|---|---|
| **Zacatecas Purple** | A type of marijuana grown in Mexico. The seeds turn purple when dried. |
| **Zig-Zag** | A brand of wheat-straw cigarette paper in which marijuana is placed for smoking. |
| **ZNA** | A mixture of dill weed and monosodium glutamate, smoked to get a minor hallucinogenic high. |
| **Zonked** | Highly intoxicated from a drug; also, sometimes, an overdose. |

# BIBLIOGRAPHY

Adriani, John. *The Chemistry and Physics of Anesthesia*, Springfield, IL: Charles C. Thomas, 1962, 849 pp.

Alexander, Deanna and O'Quinn-Larson, Josie. "When Nurses Are Addicted to Drugs, Confronting an Impaired Co-Worker", *Nursing*, Vol. 20, No. 8, August 1990, p. 55–58.

Almacan. *Drug Testing: The Supreme Court's Rulings, Federal Regulations, and Their Consequences for EAPs*, July 1989, p. 10–13.

Almacan. *The Legal Action Center Comments on the Supreme Court's Drug Testing Cases, State Legislation, and The Drug-Free Workplace Act*, July 1989, p. 10–13.

American Council For Drug Education. *Marijuana*, 5820 Hubbard Drive, Rockville, MD, 20852.

American Council For Drug Education. *Marijuana Goes to School*, 5820 Hubbard Drive, Rockville, MD, 20852.

American Council on Marijuana and Other Psychoactive Drugs. *Cocaine: Some Questions and Answers*, 6193 Executive Blvd., Rockville, MD 20852.

American Nurses' Association. *Addictions and Psychological Dysfunctions in Nursing, The Profession's Response to the Problem*.

Anonymous. "Drug Abuse in the Fire Department", *Fire Chief Magazine*, February 1985.

Anonymous. Review of: Apologia or Confessions of an English Opium Eater; Being an Extract from the Life of a Scholar. *Medico-Chirurgical Review*, Vol. 2, p. 881–901, 1822.

Anslinger, Harry J. and Tompkins, William F. *The Traffic in Narcotics*, New York: Funk and Wagnalls, 1953. 354 pp.

Archambault, R., Doran, R., Mahas, T., Nadolski, J., and Sutton-Wright, D. *Reaching Out: A Guide to EAP Casefinding*, Troy, MI: Performance Resource Press, Inc., 1982.

Arizona Criminal Justice Commission. *Arizona Drug Enforcement Strategy*, 1993.

Baldwin, DeWitt C., Jr. et al. "Substance Abuse Among Senior Medical Students", *Journal of the American Medical Association*, Vol. 265, No. 16, April 24, 1991, p. 2074–2078.

Barcroft, David, et al. *Fifty Doctors Against Alcohol*, The Brotherhood Publishing House, London, 1911.

Baywood, Threse. *Substance Abuse and Obligations to Colleagues*, Nursing Management, Vol. 21, No. 8, August 1990, p. 40–41.

Berry, C.A. *Good Health for Employees and Reduced Health Care Costs For Industry*, Washington, D.C.: Health Insurance Institute, 1981.

Blair, Brenda R. "Selecting an EAP Contractor That Will Meet Company Needs", *Occupational Health and Safety*, Nov./Dec. 1984.

Blair, Brenda R. *Supervisors and Managers Are Enablers,* Minneapolis: Johnson Institute, 1983.

Blair, T.S. "The Doctor, The Law, and the Addict", *American Medicine*, Vol. 27, 1921, p. 581–588.

Blondell, Richard D. "Impaired Physicians", *Primary Care*, Vol. 20, Number 1, March 1993.

Blum, R.H. *Society and Drugs*, San Francisco: Jossey-Bass, 1969.

Blum, R.H. *Students and Drugs*, San Francisco: Jossey-Bass, 1969.

Bracelin, Frank J. and Penna, Peter M. "Diversion of Controlled Substances", *U.S. Pharmacists*, December 1982, p. H1–H10.

Bracelin, Frank J. "Pharmacy Law", *U.S. Pharmacist*, August 1981, p 20–23.

Bray, Robert M. "Progress Toward Eliminating Drug and Alcohol Abuse among U.S. Military Personnel", *Armed Forces & Society*, Vol. 18(4), Summer, 1992, p. 476–496.

Brecher, Edward M. et al. *Licit and Illicit Drugs*, Consumer Union Report, Boston, MA: Little Brown and Company, 1972.

Brewster, Joan M. "Prevalence of Alcohol and Other Drug Problems Among Physicians", *Journal of the American Medical Association*, Vol. 255, No. 14, April 11, 1986, p. 1913–1920.

Buell, R.L. "The Opium Conferences", *Foreign Affairs*, Vol. 3, 1925, p. 567–583.

Burnett, Joseph W. "Drug Abuse", *Cutis*, Vol. 49, May 1992, p. 307–308.

Burroughs, William. *Junkie*, New York: Ace Books, 1953, 62 pp.

Business and Legal Reports, *What Everyone Should Know About Substance Abuse*, 64 Wall Street, Madison, CT.

Calabrese, Edward J., *Alcohol Interactions with Drugs and Chemicals*, Chelsea, MI: Lewis Publishers, 1991, 82 pp.

Cagney, J. Kenneth. *Beating the Drug and Alcohol Problem in the Workplace*, Madison, CT: Business and Legal Reports, 1986.

Campbell, Reginald L. "Substance Abuse — A Growing Health and Safety Problem", International Conference on the Health of Miners, *Annuals of American Conference of Governmental Industrial Hygienists*, Vol. 14, 1986, p. 489–491.

Campbell, Reginald L., Substance Abuse Takes Toll on Workers, *International Union of Operating Engineers*, Vol. 128, No. 9, Sept. 1985 p. 11–12.

Carlsen, G. "Chemical Abuse: Unfit for Duty", *Fire Engineering*, December 1984.

Castro, Janice. "Battling the Enemy Within", *Time*, March 17, 1986, p. 52–61.

Chein, I., Gerard, D.L., Lee, R.S., and Rosenfeld, E. *The Road to H: Narcotics, Delinquency, and Social Policy*, New York: Basic Books, 1964, 482 pp.

Chi, Judy and Maxwell, Tracey. "Hospitals Step Up Security to Prevent Fentanyl Abuse", *Hospital Pharmacy Report*, p. 29–30, March 1991.

Channing Bete Co., Inc. *ABC's of Drinking and Driving*, South Deerfield, MA. 01373.

Channing Bete Co., Inc. *About Drinking and Driving*, South Deerfield, MA., 01373, 1986.

Channing Bete Co., Inc. *About Antianxiety Drugs*, South Deerfield, MA., 01373, 1987.

Channing Bete Co., Inc. *About "Crack" or "Rock" Cocaine*, South Deerfield, MA., 01373, 1987.

Channing Bete Co., Inc. *About the Drug-Free Workplace Act*, South Deerfield, MA. 01373.

Channing Bete Co., Inc. *About Preventing Drug Abuse*, South Deerfield, MA. 01373.

Channing Bete Co. Inc. *About Substance Abuse At Work*, South Deerfield, MA 01373, 1987.

Channing Bete Co. Inc. *Cocaine*, South Deerfield, MA., 01373, 1985

Channing Bete Co. Inc. *Depressants,* South Deerfield, MA 01373, 1982

Channing Bete Co. Inc. *Deliriants,* South Deerfield, MA 01373, 1982

Channing Bete Co., Inc. *Drugs and You,* South Deerfield, MA., 01373, 1981.

Channing Bete Co. Inc. *Heroin,* South Deerfield, MA 01373, 1982

Channing Bete Co. Inc. *Hallucinogens,* South Deerfield, MA 01373, 1982

Channing Bete Co., Inc. *How Alcohol and Drugs Affect Your Driving Skills,* South Deerfield, MA., 01373, 1986.

Channing Bete Co. Inc. *How Alcohol and Drugs Affect Your Driving Skills,* South Deerfield, MA 01373, 1984

Channing Bete Co. Inc. *Opiates,* South Deerfield, MA 01373, 1982

Channing Bete Co. Inc., *Stimulants,* South Deerfield, MA 01373, 1982

Channing Bete Co., Inc. *What Everyone Should Know About Alcohol,* South Deerfield, MA. 01373, 1973.

Channing Bete Co., Inc. *What Everyone Should Know About Alcoholism,* South Deerfield, MA. 01373.

Channing Bete Co., Inc. *What Every Teenager Should Know About Alcohol,* South Deerfield, MA. 01373.

Channing Bete Co., Inc. *What Young Adults Should Know About Alcohol and Driving,* South Deerfield, MA. 01373.

Channing Bete Co. Inc. *What Every Parent Should Know About Drugs and Drug Abuse,* South Deerfield, MA. 01373.

Channing Bete Co., Inc. *What You Should Know About Marijuana (Military Edition),* South Deerfield, MA. 01373, 1982.

Channing Bete Co., Inc. *What You Should Know About PCP,* South Deerfield, MA, 01373, 1980

Clark, David C. et al. "Alcohol-Use Patterns Through Medical School", *Journal of the American Medical Association,* Vol. 257, No. 21, June 5, 1987, p. 2921–2926.

Clark, Mary D. "Preventing Drug Dependency, Part I, Recognizing Risk Factors", *Journal of Nursing Administration,* Vol. 18, No. 12, December 1988, p. 12–15.

Clark, Mary D. "Preventing Drug Dependency, Part II, Educating and Supporting Staff", *Journal of Nursing Administration,* Vol. 19, No. 1, January 1989, p. 21–26.

Claudio, Arnaldo. Sendero Luminos and the Narcotrafficking Alliance. *Low Intensity Conflict and Law Enforcement,* Vol. 1, No. 3, Winter, 1992, p. 379–292.

Cohen, Sidney. "Drugs in the Workplace", *Journal of Clinical Psychiatry,* December, 1984, p. 4–8.

Committee of Correspondence, Inc. *Marijuana: More Harmful Than You Think,* Danvers, MA, 1988.

Committee of Concerned Asian Scholars. *The Opium Trail,* Boston: New England Free Press, 1972.

Cornacchia, Harold et al. *Drugs in the Classroom, A Conceptual Model for School Programs,* St. Louis: C.V. Mosby Co., 1978.

Corporation Against Drug Abuse, *Employee Assistance Programs, A Resource Book for Small Employers,* Washington, DC., 1991.

Craley, D.M. "Legal Insights", *Chief Fire Executive,* May 1987.

Danne, Fredric. "Annals of Crime: Revenge of the Green Dragons", *The New Yorker,* November 16, 1992, p. 76–99.

Day, Horace (attributed to). *The Opium Habit, with Suggestions as to the Remedy,* New York: Harper & Bros., 1868.

Do It Now Publication. *Amyl/Butyl Nitrate and Nitrous Oxide,* Do It Now Foundation, P.O. Box 5115, Phoenix, AZ 85010, 1979.

Do It Now Publication. *Facts About Angel Dust,* Do It Now Foundation, P.O. Box 5115, Phoenix, AZ 85010, 1978.

Do It Now Publication. *The Chemical Eye Openers: Speedy Stuff,* Do It Now Foundation, P.O. Box 5115, Phoenix, AZ 85010, 1978.

Do It Now Publication. *Alcohol, Simple Facts About Combinations With Other Drugs,* Do It Now Foundation, P.O. Box 5115, Phoenix, AZ 85010, 1979.

Do It Now Publication. *PCP — Dream Turned Nightmare,* Do It Now Foundation, P.O. Box 5115, Phoenix, AZ 85010, 1980.

Drilling, Vern. *Alcohol and Drugs and You And Me,* Minneapolis, MN: Comp Care Publishers, 1985.

Drug Abuse Council. *The Facts About Drug Abuse,* New York: The Free Press, Macmillan Publishing Co., 1980.

Dunkin, W. *The EAP Manual,* New York: National Council on Alcoholism, 1981.

Duval, Stephen C. "Comparing Three EAPs: External Programs Are Easiest to Use", *Occupational Safety and Health,* December 1986, p. 71–73.

Eckhardt, Michael J. et al. "Health Hazards Associated With Alcohol Consumption", *Journal of the American Medical Association,* Vol. 246, No. 6, August 7, 1981, p. 648–666.

Editorial. *Journal of the American Medical Association,* Vol., 64, No. 1, 1915, p. 834–845.

Editorial, *Journal of the American Medical Association,* Vol., 64, No. 1, 1915, p. 911–912.

Ehrenfeld, Rachel. *Evil Money: Encounters along the Money Trail,* Harper Business, 1992.

Einstein, Stanley. *The Use and Misuse of Drugs,* Belmont, CA: Wadworth Publishing Co., 1970.

Estepp, M.H.,and Redding, E. "Substance Abuse in the Fire Department", *The International Fire Chief,* June 1985.

Evans, David G. "Chain-of-Custody Errors Can Quickly Undermine the Case in Court", *Occupational Safety and Health,* April 1992, p. 48–54.

Federal Emergency Management Agency (FEMA) and U.S. Fire Administration. *Administrative Manual on Use of Drugs by Fire Department Members,* FA 77, May 1988.

Fields, Albert and Tararin, Peter A. "Opium in China", *British Journal of the Addictions,* 64(3/4), 1970, p. 371–382.

Fort, J. *Pleasure Seekers. The Drug Crisis, Youth and Society,* Indianapolis: Bobbs-Merrill, 1969.

Galanter, Marc et al., "Combined Alcoholics Anonymous and Professional Care for Addicted Physicians", *American Journal of Psychiatry,* Vol. 147, No. 1, January 1990, p. 64–68.

Gelfand, Gloria, Long, Patricia, McGill, Diane, and Sheering, Cathy. "Prevention of Chemically Impaired Nursing Practice", *Nursing Management,* Vol. 21, No. 7, July 1990, p. 76–78.

Goode, Eric. *Drugs in American Society,* New York: Alfred A. Knopf, 1972.

Goldsmith, Margaret. *The Trail of Opium: The Eleventh Plague*, London: Robert Hall, Ltd., 1939, p. 286.

Gordon, Lesley. *A Country Herbal*, New York: Mayflower Books, Inc., 1980.

Green, Pat. "The Chemically Dependent Nurse", *Nursing Clinics of North America*, Vol. 24, No. 1, March 1989, p. 81–94.

Gustavsson, Nora S. "Drug Exposed Infants and their Mothers: Facts, Myths, and Needs", *Social Work in Health Care*, Vol. 16, No. 4, 1992, p. 87–100.

Halpern, S. "The Army and Heroin Addiction", *Listen, A Journal of Better Living*, August 1972.

Hawks, Richard L., and Chiang, C. Nora, eds. *Urine Drug Testing for Drugs of Abuse*, Washington, D.C.: National Institute on Drug Abuse, 1986.

Hazelden Learn About Series. *Youth and Drug Addiction*, Center City, MN: Hazelden Educational Materials, 1985.

Health Communications, Inc. *Facts About Tobacco*, 2119-A Hollywood Blvd, Hollywood, FL 33020,

Health Communications, Inc. *Facts About Inhalants*, 2119-A Hollywood Blvd, Hollywood, FL 33020.

Health Communications, Inc. *Facts About PCP, Angel Dust*, 2119-A Hollywood Blvd, Hollywood, FL 33020.

Hess, A.C. "Deviance Theory and the History of Opiates", *International Journal of the Addictions*, Vol. 7, No. 4, December 1971, p. 585, 598.

Hoffman LaRoche. *Corporate Initiatives for a Drug-Free America*, Special Report, Summer 1986.

Holiday, Billie. *An Autobiography: Lady Sings the Blues*, New York: Lancer Books, 1972.

Horman, Richard E. and Fox, Allan M., eds. *Drug Awareness: Key Documents on LSD, Marihuana, and the Drug Culture*, New York: Avon, 1969.

Howell, Sgt. William. United States Army Military Police Unit, Fort Huachuca, Arizona, personal communications.

Hughes, Patrick H. et al., "Resident Physician Substance Abuse in the United States", *Journal of the American Medical Association*, Vol. 265, No. 16, April 24, 1991, p. 2069–2073.

Iowa Board of Pharmacy Examiners, July 1985, Vol. IA 6, Number 4.

Irwin, Samuel. "Drugs of Abuse: An Introduction to their Actions and Potential Hazards", *Journal of Psychedelic Drugs*, 3(2), Spring 1971, p. 5–15.

Jaffe, Jerome H. "Drug Addiction and Drug Abuse", In: Goodman, Louis, S., and Gilman, Alfred, eds. *The Pharmacological Basis of Therapeutics*, 4th. edition, New York: Macmillan, 1970, p. 276–313.

Khantzian, Edward J. "The Injured Self, Addiction And Our Call to Duty", *Journal of the American Medical Society*, Vol. 254, No. 2, July 12, 1985, p. 249–252.

Krames Communications. *Alcoholism in the Family: Is Everyone Trapped by the Bottle?* 1992.

Krames Communications. *Alcoholism in the Workplace*, 1986.

Krames Communications. *Alcoholism in the Family: What You Can Do*, 1986.

Lewis, Walter H. and Memory, P.F. *Medical Botany, Plants Affecting Man's Health*, New York: John Wiley & Sons, 1984.

Lingeman, Richard R. *Drugs From A to Z*, New York: McGraw-Hill Paperbacks, 1974.

Lindensmith, Alfred Ray. *The Addict and the Law,* Bloomington, IN: Indiana University Press, 1965, 337 pp.

Lippman, Helen. "Addicted Nurses: Tolerated, Tormented, or Treated?", *RN,* Vol. 55, No. 4, April 1992, p. 36–41.

LoGodna, Gretchen E. and Hendrix, Melva Jo. "Impaired Nurse, A Cost Analysis", *Journal of Nursing Administration,* Vol. 19, No. 9, September 1989, p. 13–18.

Lowes, P. D. *The Genesis of International Narcotics Control,* Geneva: Librairie Droz, 1966.

Mabey, Richard, ed. *The New Age Herbalist,* New York: Collier Books, Macmillan Publishing Company, 1988.

Maddux, James F., Hoppe, Sue K., and Costello, Raymond M. "Psychoactive Substance Use Among Medical Students", *American Journal of Psychiatry,* Vol. 143, No. 2, February 1986, p. 187–191.

Mandell, Wallace et al. "Alcoholism and Occupations: A Review and Analysis of 104 Occupations", *Alcoholism: Clinical and Experimental Research,* Vol. 16, No. 4, July/August 1992, p. 734–746.

March of Dimes Birth Defects Foundation. *Teens Talk Drugs,* White Plains, NY, 1992.

Marini, Gerard. "Comprehensive Drug-Abuse Program Can Prove Effective in the Workplace", *Occupational Safety and Health,* April 1991, p. 54–59.

Martin, David W. et al. "Comprehensive Program Increases Supervisor's Knowledge of Drug Abuse", *Occupational Safety and Health,* Nov.–Dec. 1984.

McAuliffe, William E. et al. "Alcohol Use and Abuse in Random Samples of Physicians and Medical Students", *American Journal of Public Health,* Vol. 81, No. 2, February 1991, p. 177–182.

McCann, Jean. "Curbing Drug Theft", *Pharmacy World News,* August 1986, p. 12–13.

McCoy, Alfred, W. and Black, Alan A., eds. *War on Drugs: Studies in the Failure of U.S. Narcotic Policy,* Westview Press, 1992.

McEwen, J.T., Mandi, B., and Connors, E. "Employee Drug Testing Policies in Police Departments", Research in Brief, National Institute of Justice, U. S. Department of Justice, October 1986.

Mieczkowski, Tom et al. "Testing Hair for Illicit Drug Use", Research in Brief, U.S. National Institute of Justice, January 1993.

Miller, Helen. "Addiction in A Coworker, Getting Past the Denial", *American Journal of Nursing,* May 1990, pp. 72–75.

Miscall, Brian G. "Monitoring Recovering Physicians: The New Mexico Experience", *American College of Surgeons Bulletin,* Vol. 76, No. 3, March 1991, p. 22–40.

Milstead, R. *Empowering Women Alcoholics To Help Themselves and Their Sisters in The Workplace,* Dubuque, IA: Kendall-Hunt Publishing Co., 1981.

Moore, Richard D. "Youthful Precursors of Alcohol Abuse in Physicians", *The American Journal of Medicine,* Vol. 88, April 1990, p. 332–336.

Morse, Robert M. et al. "Prognosis of Physicians Treated For Alcoholism and Drug Dependence", *Journal of the American Medical Association,* Vol. 251, No. 6, Feb. 10, 1984, p. 743–746.

National Crime Prevention Council, *Don't Lose a Friend to Drugs,* 1770 K Street, Washington, D.C.

National Employee Alcoholism/Assistance Program Standards Committee. *Standards for Employee Alcoholism/Assistance Programs,* Arlington, VA: NEA/AP Standards Committee, 1982.

National Fire Protection Association. "Drug Abuse in the Fire Department", *Fire Service Section Newsletter,* Vol. 13, No. 1, 1985.

National Institute on Alcohol Abuse and Alcoholism. *Target: Alcohol Abuse in the Hard to Reach Work Force,* DHHS Pub. No. (ADM)82-1210, Washington, D.C.: U.S. Government Printing Office, 1982.

National Institute on Drug Abuse. *Employee Drug Screening — Detection Of Drug Use by Urinalysis,* Q&A, Washington, D.C.: Public Health Service, Alcohol, Drug Abuse and Mental Health Administration, 1988.

National Institute on Drug Abuse. *Hallucinogens and PCP,* Washington, D.C.: Public Health Service, Alcohol, Drug Abuse and Mental Health Administration (ADM 84-1306), 1984.

National Institute on Drug Abuse. *Inhalants,* Washington, D.C.: Public Health Service, Alcohol, Drug Abuse and Mental Health Administration (ADM 83-1307), 1983.

National Institute on Drug Abuse. *Marijuana,* Washington, D.C.: Public Health Service, Alcohol, Drug Abuse and Mental Health Administration (ADM 84-1307), 1984.

National Institute on Drug Abuse. *Medical Review Officer Manual — A Guide to Evaluating Urine Drug Screening Programs,* Washington, D.C., September, 1988.

National Institute on Drug Abuse. *Model Plan for a Comprehensive Drug-Free Workplace Program,* Washington, D.C., 1989.

National Institute on Drug Abuse. *Opiates,* Washington, D.C.: Public Health Service, Alcohol, Drug Abuse and Mental Health Administration (ADM 84-1308), 1984..

National Institute on Drug Abuse. *Sedative-Hypnotics,* Washington, D.C.: Public Health Service, Alcohol, Drug Abuse and Mental Health Administration (ADM 84-1309), 1984.

National Institute on Drug Abuse. *Stimulants and Cocaine,* Washington, D.C.: Public Health Service, Alcohol, Drug Abuse and Mental Health Administration (ADM 84-1304), 1984.

National Institute on Drug Abuse. *The Workplace as a Setting for Drug Abuse Prevention,* Washington, D.C.: Public Health Service, Alcohol, Drug Abuse and Mental Health Administration, (ADM 84-1197), 1984.

National Institute on Drug Abuse. *Workplace Drug Abuse Policy — Considerations in the Business Community,* Washington, D.C., 1989.

Nestler, Eric J. "Molecular Mechanisms of Drug Addiction", *Journal of Neuroscience,* Vol. 12, No. 7, July 1992, p. 2439–2450.

Newcomb, Michael, D. *Drug Use In the Workplace, Risk Factor for Disruptive Substance Use Among Young Adults,*

Obot, Isidore, S. "Ethical and Legal Issues in the Control of Drug Abuse and Drug Trafficking: the Nigerian Case", *Social Science and Medicine,* Vol. 35, No. 4, August, 1992, p. 481–494.

Orentlicher, David. "Drug Testing of Physicians", *Journal of the American Medical Association,* Vol. 264, No. 8, August 22/29, 1990, p. 1039–1040.

Owen, D.W. *British Opium Policy in China and India,* New Haven: Yale University Press, 1934.

Palmer, David S. *The Shining Path of Peru,* New York: St. Martin's Press, 1992.

Parkinson, R.S., ed. *Managing Health Promotion in the Workplace,* Palo Alto, CA: Mayfield Publishing Co., 1982.

Pauwels, Judith and Benzer, David G. "The Impaired Health Professional", *The Journal of Family Practice,* Vol. 29, No. 5, 1989, p. 477–484.

Pawlak, Vic. *Barbiturates: Treat or Menace?,* West Virginia State Police, Charleston, WV.

Peters, Art. "Comeback of a Child Star — Singer Frankie Lymon Returns to the Footlights after Battle with Drugs", *Ebony,* January 1967.

Pharmacists Against Drug Abuse. *The Kinds of Drugs Kids Are Getting Into,* PADA No. 1871, 1986.

Phoenix South Community Mental Health Center. *The Challenges of Change, Alarming Trends — 1991,* Phoenix: The Arizona Association of Behavioral Health Programs.

Preble, Edward and Casey, John H. "Taking Care of Business: the Heroin User's Life on the Street", *International Journal of the Addictions,* Vol. 4, 1969, p. 1–24.

Prentice, Alfred C. "The Problem of the Narcotic Drug Addict", *Journal of the American Medical Association,* Vol. 76(2), 1921, p. 1551–1554.

President's Drug Advisory Council. *Drugs Don't Work in America,* Washington, D.C.: Executive Office of the President, 1987.

Quazi, Moumin M. "Effective Drug-Free Workplace Plan Uses Worker Testing as a Deterrent", *Occupational Health and Safety,* June 1993, p. 26–32.

Ratner, Richard A. "Drugs and Despair in Vietnam", *University of Chicago Magazine,* May/June 1972, p. 15–23.

Ray, Oakley S. *Drugs, Society and Human Behavior,* St. Louis: C.V. Mosby Co., 1974.

Reuter, Peter. *Quest for Integrity: the Mexican-U.S. Drug Issues in the 1980's,* Rand, 1992.

Rogowski, Jeannette A. "Insurance Coverage for Drug Abuse", *Health Affairs,* Vol. 11, No. 3, Autumn 1992, p. 137–148.

Romberg, Robert W. "Factors Influencing Confirmed Drug Positives for Navy and Marine Corps Recruits", *Military Medicine,* Vol. 157, 1992, p. 33–37.

Rush, Harold M.F. and Brown, James K. "The Drug Problem In Business", *The Conference Board Record,* March 1971.

Segalla, E. *Employee Assistance Programs For Local Governments,* Management Information Service Report 14(8), Washington, D.C.: International City Management Association, August 1982.

Schultes, R.E. and Hogmann, A. *The Botany and Chemistry of Hallucinogens,* Springfield, IL: Charles C Thomas, 1973.

Schwartz, Ronald M. "Drug Diversion Grows", *American Druggist,* p. 43–44, September 1990.

Shirley, Charles, E. "Alcoholism and Drug Abuse in the Workplace", *Office Administration and Automation,* November 1984, p. 24–27.

Shore, James, H. "The Oregon Experience With Impaired Physicians on Probation", *Journal of the American Medical Association,* Vol. 257, No. 21, June 5, 1987, p. 2931–2934.

Smith, D.E. and Wesson, D.R., eds. *Uppers and Downers,* Englewood Cliffs, N.J.: Prentice-Hall, Inc., 1973.

Smith, David E. and Gay, George R., eds. *It's So Good, Don't Even Try It Once: Heroin in Perspective,* Englewood Cliffs, N.J.: Prentice-Hall, 1972, 208 pp.

Smith, Douglas A. "Specifying the Relationship between Arrestee Drug Test Results and Recidivism", *The Journal of Criminal Law and Criminology,* Vol. 83, No. 2, 1992, p. 364–377.

Smith, R. Blake. "Information — a Critical Resource In Fight Against Substance Abuse", *Occupational Safety and Health,* April 1992, p. 57.

Sonndecker, Glenn. "Emergence and Concept of the Addiction Problem", Symposium on the History of Narcotics Addiction, Bethesda, MD: U.S. Public Health Service, 1958, p. 14–22.

Stahl, Marc B. "Asset Forfeiture, Burdens of Proof, and the War on Drugs", *The Journal of Criminal Law and Criminology,* Vol. 83, No. 2, 1992, p. 274–327.

Stewart, W. Wayne, ed. *Drug Abuse In Industry,* Miami: Halos, 1970.

Strang, J. "Past and Present Perspectives on Cocaine in Britain", *Verhandlelingen-Koninklijke Academie voor Geneeskunda von Belgie,* Vol. 54, No. 5, 1991, p. 547–555.

Sudduth, Ann B. "EAPs: Defining Functions and Evaluating Your Program", *Occupational Health and Safety,* Nov./Dec. 1984, p. 44–53.

Superweed, Mary Jane. *Herbal: A Guide to Natural and Legal Narcotics, Psychedelics and Stimulants,* Stone Kingdom Syndicate.

Svenson, James. "A Physician's Dilemma", *Journal of the American Medical Association,* Vol. 259, No. 18, May 13, 1988, p. 2749.

Talbott, C. Douglas et al. "The Medical Association of Georgia's Impaired Physicians Program", *Journal of the American Medical Association,* Vol. 257, No. 21, June 5, 1987, p. 2927–2930.

Tate, Cassandra. "In the 1800's Antismoking Was a Burning Issue", *Smithsonian,* Vol. 20, No. 4, July 1989.

Taylor, Douglas N. "Self-Esteem, Anxiety, and Drug Abuse", *Psychological Reports,* Vol. 71, 1992, p. 896–898.

Terry, Charles E. and Pellens, Mildred. *The Opium Problem,* New York: Bureau of Social Hygiene, Inc., 1928, 1042 pp.

Towns, Charles B. *The Necessity of Definite Medical Result In The Treatment of Drug and Alcohol Addiction,* New York: Charles B. Towns Hospital, 1917.

Tully, Andrew. *The Secret War Against Dope,* New York: Coward, McCann & Geoghegan, Inc., 1973.

Uchida, Craig, D. *Modern Policing and the Control of Illegal Drugs.* U.S. Department of Justice, Office of Justice Programs, National Institute of Justice: Police Foundation, 1992.

Ulwelling, John J. "The Evolution of the Oregon Program for Impaired Physicians", *American College of Surgeons Bulletin,* Vol. 76, No. 3, March 1991, p. 18–21.

Ukens, Carol. "Health Care's Addicts Deserve Patient Care, Too", Hospital Pharmacist Report, January 1991, p. 23.

United Nations, International Narcotics Control Board Report. *View of the Board on the Question of Legalization of the Non-Medical Use of Drugs,* 1992, p. 3–6.

United States of America, Executive Office of the President, President's Drug Advisory Council. *Drugs Don't Work In America,* Washington, DC.

United States Department of Health and Human Services. *Drugs in the Workplace, Research and Evaluation Data,* Research Monograph Series, Washington, D.C.,

United States Department of Health and Human Services. *Hallucinogens and PCP,* Public Health Service, Alcohol, Drug Abuse and Mental Health Administration, DHHS Pub. No. (ADM 83-1306), Washington, DC., 1983.

United States Department of Health and Human Services. *Cocaine: Summaries of Psychosocial Research,* Research and Evaluation Data, Research Monograph Series, No. 15 (ADM 84-391), 1984.

United States Department of Health and Human Services. *Cocaine: Pharmacology, Effects, and Treatment of Abuse,* Research and Evaluation Data, Research Monograph Series, No. 50 (ADM 84-1326), 1984.

United States Department of Health and Human Services. *Marijuana Effects on the Endocrine and Reproductive Systems,* Research and Evaluation Data, Research Monograph Series, No. 44 (ADM 84-1278), 1984.

United States Department of Health and Human Services. *For Parents Only: What You Need to Know about Marijuana* (ADM 81-909), 1981.

United States Department of Health and Human Services. *Research on Drugs and the Workplace,* NIDA Capsules, C-87-2, revised 06/1990.

United States Department of Health and Human Services. *Let's Talk About Drug Abuse,* (ADM 81-706), 1981.

United States Department of Health and Human Services. *Mandatory Guidelines For Federal Drug Testing Programs,* NIDA Capsules, C-88-01, April 1988.

United States Department of Health and Human Services. *Summary of Findings from the 1991 National Household Survey on Drug Abuse,* NIDA Capsules, C-86-13, revised December 1991

United States Department of Health and Human Services. *Comprehensive Procedures For Drug Testing In the Workplace,* NIDA (ADM 91-1731), 1991.

United States Department of Health and Human Services. *Employee Drug Screening,* NIDA (ADM 88-1442), revised 1988.

United States Department of Health and Human Services. *Communities: What You Can Do About Drug And Alcohol Abuse,* NIDA (ADM 84-1310), 1984.

United States Department of Health and Human Services. *Elder-Ed Group Leader's Guide to Wise Use of Drugs, A Program for Older Americans,* NIDA (ADM 84-1074), 1984.

United States Department of Health and Human Services. *Elder-Ed: Using Your Medicines Wisely: A Guide For the Elderly,* NIDA (ADM 84-705), 1984.

United States Department of Health and Human Services. *Employer's Guide to the Employment of Former Drug and Alcohol Abusers,* NIDA (ADM 83-1292), 1983.

United States Department of Health and Human Services. *The Drinking Question - Honest Answers to Questions Teenagers Ask About Drinking,* NIDA (ADM 76-286), 1976.

United States Department of Health and Human Services. *Here Are Some Things You Should Know About Prescription Drugs,* FDA (FDA 84-3124), 1984.

United States Department of Health and Human Services. *Drugs in the Workplace,* Research and Evaluation Data, Vol. II, Research Monograph 100, 1990.

U.S. Department of Health and Human Services. *Mandatory Guidelines for Federal Workplace Drug Testing Programs. Federal Register,* Part IV, Monday, April 11, 1988.

United States Department of Health and Human Services. *Parents, Peers and Pot,* DHHS Publication (ADM 80-812), 1980.

United States Department of Justice, Drug Enforcement Administration, Office of Intelligence, Strategic Intelligence Section, Domestic Unit. *Illegal Drug Price/Purity Report, 1988 through September 1991,* January, 1992.

United States Department of Justice, Drug Enforcement Administration. *Drug Abuse and Misuse,* GPO 1979 O-293-289, Washington, D.C.: U.S. Government Printing Office, 1979.

United States Department of Justice, Drug Enforcement Administration. *Drugs of Abuse,* Vol. 6, No. 2, GPO 1985-482-711, Washington, D.C.: U.S. Government Printing Office, 1985.

United States Department of Justice, Drug Enforcement Administration. *"Drug Abuse In the Workplace",* Public Affairs Fact Sheet.

United States Department of Justice, Drug Enforcement Administration. *Controlled Substances: Use, Abuse and Effects,* GPO 1981 O-351-534, Washington, D.C.: U.S. Government Printing Office, 1981.

United States Department of Labor, Mine Safety and Health Administration. "Alcoholism and Drug Abuse — How Prevalent in Mining?", Quarterly Magazine, 1990.

United States Department of Transportation, National Highway Traffic Safety Administration. *How to Talk to Your Teenager About Drinking and Driving,* Washington, D.C., October, 1975.

Up Front, Inc. *Up Front About Barbiturates,* Box 330589, Coconut Grove, Fl. 33133.

Up Front, Inc. *Up Front About Cocaine,* Box 330589, Coconut Grove, FL, 33133

Usdin, E. and Efron, D.H. *Psychotropic Drugs and Related Compounds,* Washington, D.C.: U.S. Government Printing Office, 1972.

Vicary, J.R. and Resnick, H. *Preventing Drug Abuse in the Workplace,* DHHS Pub. No. ADM 82-1220, Washington, D.C.: U.S. Government Printing Office, 1982.

Wagel, William H. "A Drug Screening Policy That Safeguards Employees' Rights", *Personnel,* February 1988, p. 10–11.

Walsh, J.M. and S.W. Gust. "Drug Abuse in the Workplace". *Seminars in Occupational Medicine,* Vol. 1, No. 4, December 1986.

Walsh, J. Michael. "Drug Testing in the Private and Public Sectors", *Bulletin of the New York Academy of Medicine,* Vol. 65, No. 2, February 1989, p. 166–172.

Webster, Barbara. "International Money Laundering: Research and Investigation Join Forces", *U.S. National Institute of Justice Research in Brief,* September, 1992.

Weisner, Constance. "The Merging of Alcohol and Drug Treatment: A Policy Review", Journal *of Public Health Policy,* Vol. 13, No. 1, Spring, 1992, p. 66–80.

West Virginia State Police. *Angel Dust, Devil In Disguise,* Charleston, WV.

West Virginia State Police. *Quaaludes and Other Sopors,* Charleston, WV.

White, John P. "Hospitals Seen As Narcotic 'Supermarket' for Employees", *Drug Topics,* Jan. 1986, p. 79.

Williams, Etta. "Strategies for Intervention", *Nursing Clinics of North America,* Vol. 24, No. 2, March 1989, p. 95–107.

Wood, Horatio C. *A Treatise on Therapeutics and Pharmacolog Materia Medica,* 2 vols., Philadelphia: Lippincott, 1884.

Young, George. *A Treatise on Opium,* London: A. Miller, 1753, 182 pp.

Zarafonetis, D.J.D., ed. *Drug Abuse: Proceedings of the International Conference, Ann Arbor,* 1970, Philadelphia: Lea and Febiger, 1972.

Zinberg, Norman E. "Heroin Use in Vietnam and the United States", *Archives of General Psychiatry,* Vol. 26, No. 5, May 1972, p. 486–488.

# INDEX

## A

Absinthe, in history of drug abuse, 12
Accidents, and workplace substance abuse, 112
Acetaminophen, chemistry of, 41
Actidil®, chemistry of, 45
Acute, defined, 51
Acute exposure, toxicology of, 52
Adrenals, and drug abuse, 22
Aerosol propellants, chemistry of, 48
Airline pilots, substance abuse, 102
Alcohol, 26
    abuse case study, 115, 116, 118, 119, 120, 121, 122
    chemistry of, 30, 31, 34, 50
    and employee assistance program, 79
    in history of drug abuse, 8, 10
    interaction with chemicals, 46, 64, 66
    legislation against, 97
    workplace abuse, 102, 103, 105, 106, 108, 109, 110, 111, 112
Alfentanil®, chemistry of, 37
Alkaloids, chemistry of, 34
Alveoli, and drug abuse, 17
Aminopterin, toxicology of, 58
Amphetamines
    chemistry of, 30, 33
    in history of drug abuse, 9
    legislation against, 97
    toxicology of, 58
    workplace abuse, 103
Amyl nitrate, chemistry of, 48
Analgesics, chemistry of, 30, 40
Anatomy, and drug abuse, 15
Anesthetics
    chemistry of, 30, 32
    toxicology of, 53
Angel dust, chemistry of, 39, 48
Anorexics, chemistry of, 30
Antibiotics, chemistry of, 30
Antidepressants, chemistry of, 30
Antifreeze, chemistry of, 48
Antihistamines, chemistry of, 44
Antiinflammatory agents, chemistry of, 31

Antiphychotics, chemistry of, 30
Aphrodisiac, chemistry of, 29
Arteries, and drug abuse, 24
Aspartame (NutraSweet®), chemistry of, 31
Asphyxiants, toxicology of, 53
Aspirin
    chemistry of, 40
    in history of drug abuse, 5
    substitutes, chemistry of, 41
    toxicology of, 58
Atria, and drug abuse, 16
Atropine
    chemistry of, 35
    in history of drug abuse, 7

## B

Barbital, legislation against, 96
Barbiturates
    chemistry of, 30, 32, 33
    legislation against, 97
    toxicology of, 58
Barbituric acid
    chemistry of, 33
    legislation against, 96
Beer, workplace abuse, 108
Benadryl®, chemistry of, 45
Benzene, interaction with chemicals, 64
Benzo(a)pyrene, interaction with chemicals, 65
Benzodiazepines, chemistry of, 30, 34
Beta blockers, and drug abuse, 24
Bile, and drug abuse, 25
Birth defects
    chemistry of, 34, 43
    toxicology of, 58
Bladder, and drug abuse, 25
Blood, and drug abuse, 22
Body, and drug abuse, 15–26
Boggs Act of 1951, 93
Brain, and drug abuse, 19
Brevital®, chemistry of, 33
Bromides, chemistry of, 45
Bronchi, and drug abuse, 16
Bronchial, dilators, chemistry of, 31
Bronchioles, and drug abuse, 16

Printed in the United States
by Baker & Taylor Publisher Services